REVISE OCR AS/A LEVEL

Physics

REVISION

Series Consultant: Harry Smith

Authors: Steve Adams and John Balcombe

Our revision resources are the smart choice for those revising for OCR AS/A Level Physics. This book will help you to:

- **Organise** your revision with the one-topic-per-page format
- **Prepare** for your AS/A Level exam with a book full of exam-style practice questions
- **Simplify** your revision by writing straight into the book just as you would in an exam
- **Track** your progress with at-a-glance check boxes
- **Improve** your understanding, and exam technique, with guided questions to build confidence, and hints to support key revision points.

Revision is more than just this Workbook!

Make sure that you have practised every topic covered in this book, with the accompanying OCR AS/A Level Physics Revision Guide. It gives you:

- A 1-to-1 page match with this Workbook
- Explanations of key concepts delivered in short memorable chunks
- Key hints and tips to reinforc
- Worked examples showing yo ers
- Exam-style practice questions with answers.

For the full range of Pearson revision titles across KS2, KS3, GCSE, AS/A Level and BTEC visit:
www.pearsonschools.co.uk/revise

Contents

AS & A level

A level

Quantities and units

1 Which of the following is **not** an SI base quantity?

☐ **A** mass ☐ **B** length ☐ **C** charge ☐ **D** time **(1 mark)**

2 In which of the following derived units can mechanical work be expressed?

☐ **A** $kg\,m\,s^{-2}$

☐ **B** $kg\,m\,s^{-1}$

☐ **C** $kg^2\,m\,s^{-2}$

☐ **D** $kg\,m^2\,s^{-2}$

Remember that the defining equation of work is work = force × displacement. The units must reflect this.

(1 mark)

Guided

3 Link the derived units on the right with their more familiar equivalents on the left.

newton	$kg\,m^2\,s^{-3}\,A^{-1}$
volt	$kg\,m^2\,s^{-3}\,A^{-2}$
watt	$kg\,m\,s^{-2}$
ohm	$kg\,m^2\,s^{-3}$

The defining equation for force is $F = ma$, so the unit is $kg\,m\,s^{-2}$

(3 marks)

4 Given that the resistance R of a wire of length l, cross-sectional area A and resistivity ρ is given by $R = \frac{\rho l}{A}$:

(a) show that the units of resistivity are $\Omega\,m$.

Guided

Rearranging the equation gives $\rho = \frac{RA}{l}$. Considering the units of the quantities on the right-hand side gives ....

....

.... **(2 marks)**

(b) Use the above equation to show that the resistance of 1.5 km of copper wire of cross-sectional area 5.0 mm² is about 5 Ω. (The resistivity of copper is $1.68 \times 10^{-8}\,\Omega\,m$.)

....

....

....

.... **(3 marks)**

5 The speed limit on French highways is $130\,km\,h^{-1}$. Convert this speed limit to $m\,s^{-1}$.

....

....

.... **(2 marks)**

Had a go ☐ Nearly there ☐ Nailed it! ☐

Estimating physical quantities

1 The wavelength of a green laser pointer is 532 nm. This could also be written as:

☐ **A** 0.0532 mm

☐ **B** 0.532 μm

☐ **C** 0.00000532 m

☐ **D** 5.32×10^{-9} m

(1 mark)

2 Show that a year is about 32 megaseconds.

....

....

.... **(2 marks)**

Guided **3** Estimate the average electrical power required by a small town.

Method: Let's assume the town has 20 000 residents in 5000 homes plus shops, factories and schools, etc. There will be periods of high demand, e.g. evenings, and low demand, e.g. at night. ....

....

....

....

....

.... **(3 marks)**

4 Estimate the force on your legs if you jump off a 2 m high wall and land on your feet.

....

....

....

....

.... **(3 marks)**

5 It is said that the beam of the Large Hadron Collider (LHC) at CERN has the energy of an express train. If the beam at the LHC has an energy of 360 MJ, is the claim is reasonable?

Start by estimating the mass of the train by comparing it with something more familiar like a car that has a mass of about 1500 kg.

....

....

....

....

....

.... **(2 marks)**

Experimental measurements

1 A student measures the diameter of a glass rod 5 times with a micrometer and obtains the results: 5.11 mm, 5.13 mm, 5.10 mm, 5.14 mm, 5.13 mm. The precision of the micrometer measurements is 0.01 mm. The diameter of the rod should be written as:

☐ **A** 5.12 ± 0.01 mm

☐ **B** 5.13 ± 0.01 mm

☐ **C** 5.122 ± 0.02 mm

☐ **D** 5.12 ± 0.02 mm

(1 mark)

Remember that calculating a mean or average does not improve the precision of the quantity (the number of decimal places).

2 This question is about random and systematic errors in measurements.

(a) Define 'error' in an experimental measurement.

....

.... **(1 mark)**

(b) What is the difference between a random error and a systematic error?

....

....

.... **(2 marks)**

Guided

(c) Explain why taking an average of several readings will reduce random error but will not reduce systematic error.

A random error has an equal chance of being positive or negative, so

taking an average will ....

....

....

.... **(2 marks)**

(d) How might plotting a graph of one quantity against another reveal the existence of a systematic error in the measurement of those quantities?

....

.... **(1 mark)**

3 100 g masses are calibrated to an accuracy of ±5 g. Ten of them are placed on an accurate electronic balance. The reading on the balance will most likely be closest to:

☐ **A** 1000 g

☐ **B** 1050 g

☐ **C** 1100 g

☐ **D** 950 g

(1 mark)

Had a go ☐ Nearly there ☐ Nailed it! ☐

Combining errors

1 The voltmeter in this circuit is calibrated to an accuracy of 5%. When used to measure the potential difference across the thermistor, it reads 4.7 V.

6 V
A
100 Ω
V

(a) Explain the term 'calibrated'.

....

....

.... **(1 mark)**

(b) Calculate the absolute uncertainty in this measurement.

....

.... **(2 marks)**

(c) The digital ammeter reads 0.010 A. Its maximum reading is 1.999 A with a resolution of 0.001 A. Explain the term 'resolution'.

....

.... **(1 mark)**

Guided

(d) Explain why the uncertainty in the current measurement is **at least** 10%.

The reading can only be to the nearest 0.001 A, so when the current being measured is 0.010 A ....

....

.... **(3 marks)**

The resistance of the thermistor can be calculated using $R = \frac{V}{I}$.

(e) Determine the value of the thermistor's resistance.

....

....

.... **(1 mark)**

(f) Determine the percentage uncertainty in the resistance of the thermistor.

.... **(1 mark)**

(g) Determine the absolute uncertainty in the resistance of the thermistor.

.... **(1 mark)**

2 The thickness of a stack of 50 sheets of paper is measured with callipers and is found to be 6.5 ± 0.1 mm. The thickness of a single sheet of the paper is best written as:

☐ **A** 0.13 ± 0.1 mm

☐ **B** 0.13 ± 0.002 mm

☐ **C** 0.130 ± 0.002 mm

☐ **D** 0.130 ± 0.1 mm **(1 mark)**

Graphs

1 A coil of copper wire is heated and its resistance is measured at various temperatures. The results are shown in the table below.

Temperature / °C	Resistance / Ω
0	10.1
20	10.7
40	11.7
60	12.6
80	13.4
100	13.9

The temperature was measured with a negligible uncertainty. However, the resistance was measured with an uncertainty of ±0.2 Ω.

(a) Using the axes below, plot a graph of resistance against temperature for the copper coil, adding a best-fit line and error bars as appropriate. **(6 marks)**

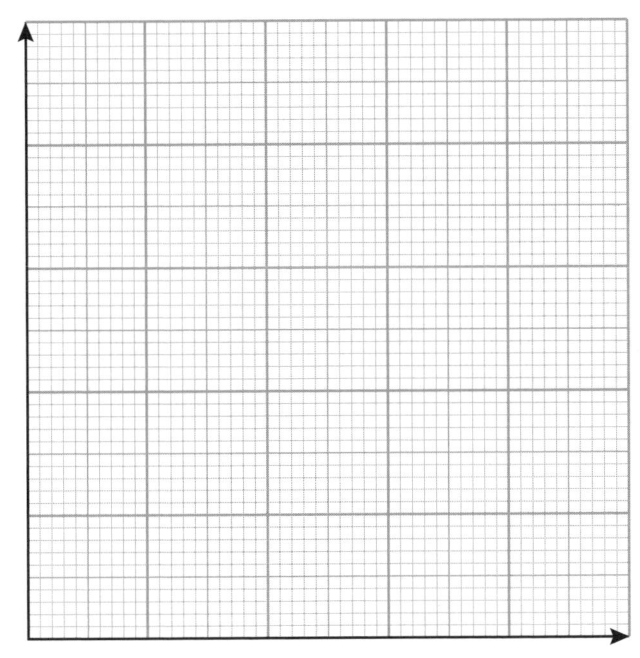

When plotting a graph, the phrase is: plot the 'vertical quantity' against the 'horizontal quantity'. The horizontal quantity should be the independent variable. Examine the range of each quantity and choose appropriate scales to maximise the use of the grid provided.

(b) Determine the gradient of the line and state its value with appropriate units.

....

....

.... **(2 marks)**

(c) Use your answer to part (b) to predict the resistance of the coil at 500 °C.

....

....

.... **(2 marks)**

(d) Suggest why your answer to part (c) might not be valid.

....

.... **(2 marks)**

Had a go ☐ Nearly there ☐ Nailed it! ☐

Scalars and vectors

1 The quantities in the table below are either scalars or vectors. Indicate which is which by a tick in the appropriate column.

Quantity	Vector	Scalar
Distance		
Momentum		
Speed		
Pressure		

(4 marks)

Guided

2 (a) Explain why adding together two particular scalar quantities always produces the same result whereas adding two vectors with the same magnitudes may produce different results.

Scalar quantities are just magnitudes, so adding them up can only produce one answer, but vectors ....

....

.... **(2 marks)**

Guided

(b) Two forces of magnitude 30 N and 40 N act on a body. Which of the following could equal the magnitude of the resultant force acting on the body? Circle the correct answer(s).

The forces could be in the same direction, when they add up to 70 N, or in opposite directions, when the resultant is 10 N, or they could add up to anything in between when they are at an angle to each other.

0 N   10 N   30 N   70 N   80 N **(2 marks)**

(c) Which of the following statements A–D concerning displacement is correct?

☐ **A** displacement = velocity × time, and time is a scalar quantity

☐ **B** displacement = velocity × time, and time is a vector quantity

☐ **C** displacement = speed × time, and speed is a vector quantity

☐ **D** displacement = speed × time, and speed is a scalar quantity. **(1 mark)**

3 Berhanu argues that temperature is a vector because it can go up or down. Cathy disagrees. Who is right and why?

....

....

....

....

.... **(2 marks)**

Vector triangles

1 The figure shows a car parked on a steep slope. Three forces act on the car. The weight and the normal reaction force are known.

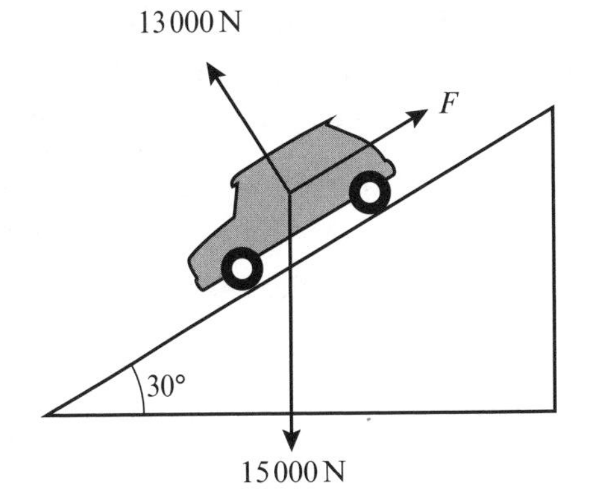

(a) Draw, to an appropriate scale, a vector triangle that indicates that these three forces are in equilibrium.

A scale diagram must be accurate in terms of length and direction.

(3 marks)

(b) How does your diagram shows that the three forces are in equilibrium?

....

....

.... **(2 marks)**

(c) Use your diagram to determine the magnitude of the frictional force, F.

....

.... **(1 mark)**

2 Two walkers at position A encounter a patch of boggy ground (map in figure right). They can walk 5.0 km due north from A to B then 3.0 km due east from B to C thus avoiding the wet ground.

B• C•

A•

(a) How far is it from A to C in a straight line?

....

....

.... **(2 marks)**

(b) The walkers can walk at $4.0\,\text{km}\,\text{h}^{-1}$ on dry ground but can only walk at $3.0\,\text{km}\,\text{h}^{-1}$ across the bog. How much time would they save taking the direct route?

....

....

....

.... **(2 marks)**

Had a go ☐ Nearly there ☐ Nailed it! ☐

Resolving vectors

1 A wrecking ball is a heavy steel mass on the end of a wire cable. It can be swung into a building that is to be demolished. The ball is pulled back by a horizontal force F and then released. The weight of the ball is 20 000 N.

(a) Add arrows to the diagram to indicate the direction of the weight (W) of the ball and the tension (T) in the wire cable.

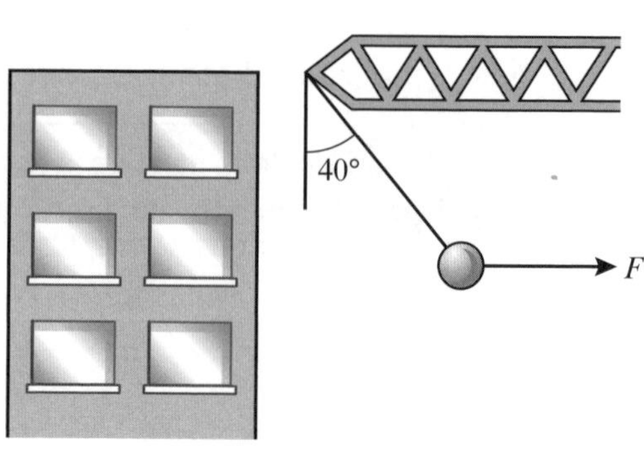

(2 marks)

(b) Determine the tension in the wire cable.

Equate the weight with the vertical component of the tension.

....

....

.... **(3 marks)**

(c) Determine the magnitude of the force F.

....

....

.... **(2 marks)**

(d) Determine the initial acceleration of the ball in the direction at right angles to the wire cable.

Consider the component of the weight of the ball at right angles to the cable. You will need g from the data sheet.

....

....

.... **(2 marks)**

2 A box of weight W is pushed up a slope of angle θ against a frictional force F (see figure). The force up the slope required to move the box at constant speed is:

Considering forces parallel to the slope, there is friction and the component of the weight, $W \sin \theta$

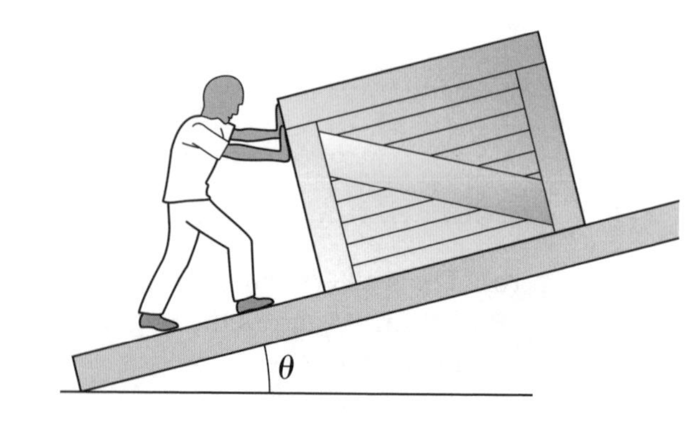

☐ **A** $W + F$

☐ **B** $F + W \sin \theta$

☐ **C** $F + W \cos \theta$

☐ **D** $W + F \sin \theta$

(1 mark)

Describing motion

1 Define the following terms used to describe motion.

(a) Displacement

.... **(1 mark)**

(b) Velocity

.... **(1 mark)**

(c) Acceleration

.... **(1 mark)**

2 A sports car accelerates in a straight line from rest to a speed of $30\,\text{m}\,\text{s}^{-1}$ in a time of 6 s. While accelerating, it covers a distance of 120 m.

(a) Determine the average speed of the car while it is accelerating.

....

.... **(2 marks)**

Guided

(b) How does your result for part (a) show that the car is not accelerating uniformly?

If an object is accelerating uniformly, its final speed must be twice its average speed, but in this case ....

....

.... **(2 marks)**

(c) Determine the average acceleration of the car over the 6 s.

....

.... **(2 marks)**

(d) Over the next 8 seconds, the car accelerates uniformly up to a speed of $40\,\text{m}\,\text{s}^{-1}$. What is the instantaneous speed of the car 10 s after it started from rest?

....

....

.... **(2 marks)**

3 An object moving at speed u accelerates uniformly to speed v in a time t. Which of the following relationships is valid?

☐ **A** average speed $= \frac{(v-u)}{2}$

☐ **B** average acceleration $= \frac{(u+v)}{2t}$

☐ **C** average acceleration $= \frac{(v-u)}{2t}$

☐ **D** average speed $= \frac{(u+v)}{2}$

(1 mark)

Had a go ☐ Nearly there ☐ Nailed it! ☐

Graphs of motion

1 The **velocity** of a car is described by the graph below.

Look carefully at the axes of the v–t graph.

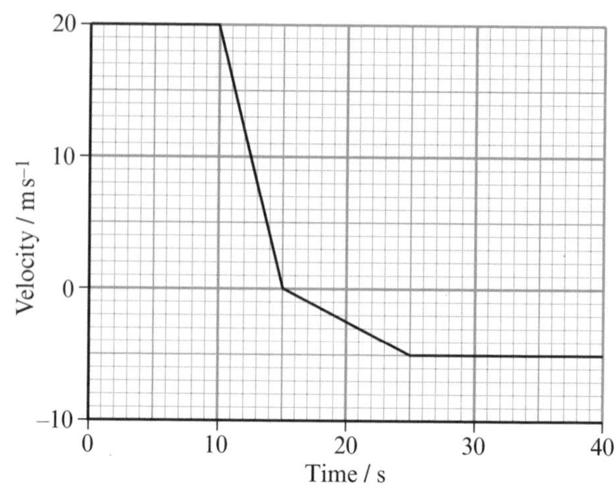

Guided

(a) Using the grid to the right, sketch a **displacement**/time graph that describes the motion of the car from 0 to 40 s.

Remember that the area under the line of a v–t graph is the displacement.

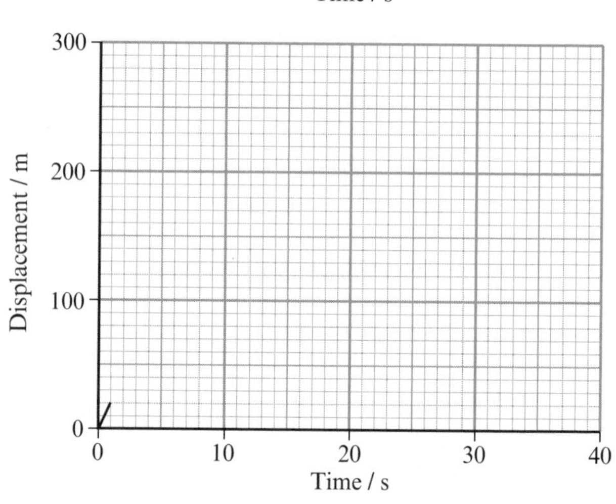

(4 marks)

(b) Using the grid to the right, sketch an **acceleration**/time graph that describes the motion of the car from 0 to 40 s.

Remember that the gradient of the line of a v–t graph is the acceleration.

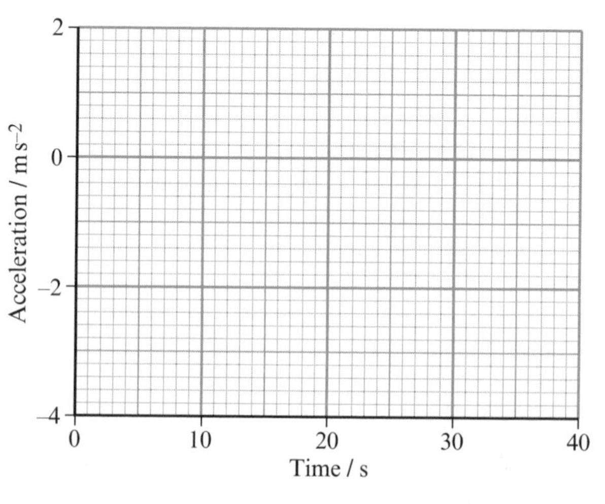

(4 marks)

(c) Calculate the total distance that the car travels from 0 to 40 s.

....

....

.... **(2 marks)**

SUVAT equations of motion

1 Two cars, each 5.0 m long, are driven at the same constant velocity of 25.0 m s^{-12} in adjacent lanes on a straight road as shown in the figure.

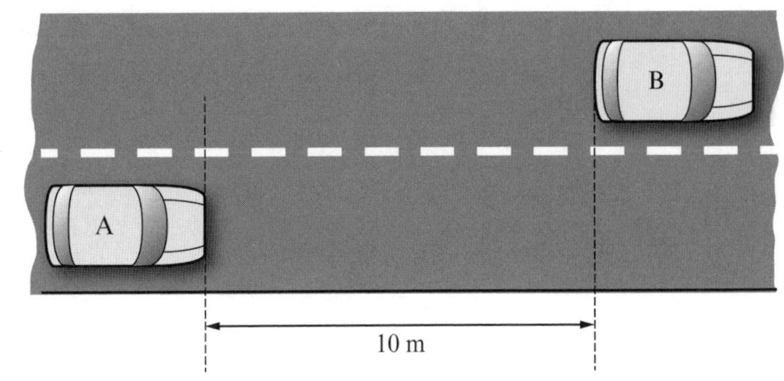

(a) The driver of car A then decides to overtake car B and accelerates uniformly at 2.0 m s^{-2}. Show that it will take about 6 s until the rear of car A is 15.0 m ahead of the front of car B.

If the time is t, B travels $25t$ metres. A (using $s = ut + \frac{1}{2}at^2$) travels $25t + \frac{1}{2} \times 2 \times t^2$ metres. A must travel $10 + 15 + 5 + 5 = 35$ m further than B ....

....

....

.... **(3 marks)**

(b) How far will car B travel during this manoeuvre?

....

....

.... **(1 mark)**

(c) How far will car A travel during this manoeuvre?

....

....

.... **(1 mark)**

(d) Determine the final speed of car A after the manoeuvre.

....

....

.... **(2 marks)**

2 Show that the equation $s = ut + \frac{1}{2}at^2$ is **homogeneous**.

....

....

....

.... **(3 marks)**

Had a go ☐ Nearly there ☐ Nailed it! ☐

Acceleration of free fall

1 Explain why, in the absence of air resistance, all objects, irrespective of their masses, would be expected to accelerate at the same rate in a uniform gravitational field, for example near the surface of the Earth.

Consider the force that makes a falling object accelerate and Newton's second law of motion.

....

....

....

.... **(4 marks)**

2 A coin is dropped down a deep well and is heard to hit the bottom 2 seconds later.

Guided (a) Estimate the depth of the well to an appropriate precision.

Using $s = ut + \frac{1}{2}at^2$ ....

....

.... **(2 marks)**

(b) The speed of sound in air is about $340\,\text{m}\,\text{s}^{-1}$. Explain why the speed of sound does not need to be taken into account in part (a) above.

....

....

.... **(2 marks)**

Guided **3** A flea can jump to a height of 0.15 m on the Earth where $g = 9.81\,\text{N}\,\text{kg}^{-1}$. On the Moon, $g = 1.62\,\text{N}\,\text{kg}^{-1}$.

(a) Show that the vertical take-off speed of a flea is about $2\,\text{m}\,\text{s}^{-1}$.

Using $v^2 - u^2 = 2as$ ....

....

.... **(2 marks)**

(b) How high could a flea jump if just the Moon's gravitational force were acting on it?

....

.... **(2 marks)**

(c) A flea achieves take-off speed over a distance of 0.8 mm. How many times its own weight would a flea experience during take-off? State any assumptions you make.

....

....

.... **(3 marks)**

Vehicle stopping distances

1 According to the Highway Code, the stopping distance of a car driven at 48 $km\,h^{-1}$ is 23 m. The likely stopping distance at 96 $km\,h^{-1}$ is:

☐ **A** 36 m ☐ **B** 46 m ☐ **C** 73 m ☐ **D** 92 m **(1 mark)**

2 The speed/time graph in the figure below describes the motion of a car that stops as a result of the driver observing a hazard. The initial speed of the car is 20 $m\,s^{-1}$.

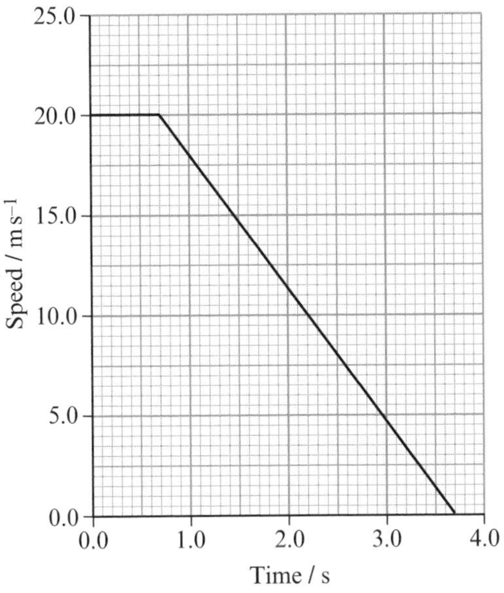

(a) State the **thinking time** of the driver.

.... **(1 mark)**

Guided

(b) Determine the **thinking distance**.

Thinking distance is the area under the horizontal part of the line, which is ....

.... **(2 marks)**

(c) Determine the **braking distance**.

....

.... **(2 marks)**

(d) Determine the total **stopping distance** at 20 $m\,s^{-1}$.

.... **(1 mark)**

(e) Add a speed/time plot for a car travelling at an initial speed of 15 $m\,s^{-1}$ to the above graph and hence determine the total **stopping distance** at 15 $m\,s^{-1}$.

> The reduced initial speed will not affect the deceleration, that is, the slope of the braking part of the graph.

....

....

....

.... **(4 marks)**

Had a go ☐ Nearly there ☐ Nailed it! ☐

Projectile motion

1 A 19th century cannon could fire a cannonball with a velocity of 480 $m s^{-1}$. When mounted high up on a cliff-top, this gave it considerable range. A ball is fired horizontally from such a cannon, and descends a total distance of 80 m during its flight.

You invariably need to consider the vertical motion first in projectile questions.

Guided

(a) Ignoring air resistance, determine the time of flight of the ball.

Using $s = ut + \frac{1}{2}at^2$ ....

.... **(2 marks)**

(b) Determine the horizontal range of the cannon.

....

.... **(2 marks)**

(c) On the axes, sketch a graph of the trajectory of the ball as calculated. **(4 marks)**

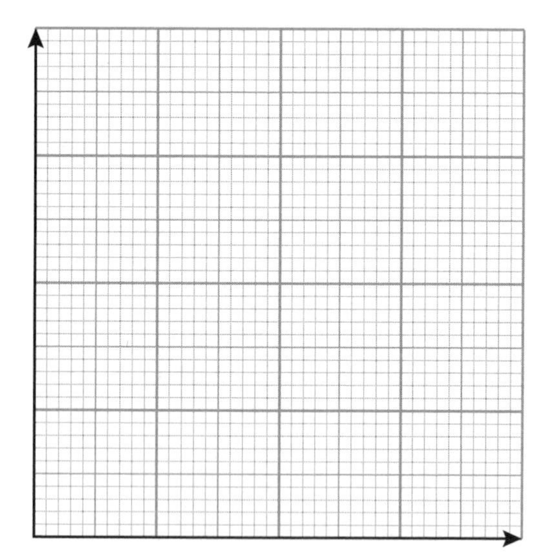

(d) Add a line to the above graph to indicate the effect of air resistance on the trajectory. **(2 marks)**

2 When a projectile is launched at an angle θ to the horizontal with initial velocity u its range r is given by:

r will be a maximum when $\sin 2\theta$ is a maximum.

$$r = \frac{u^2 \sin 2\theta}{g}$$

Explain why this formula predicts a maximum range when θ is equal to 45°.

....

....

.... **(2 marks)**

Types of force

1 Isaac is in a **moving** lift (see figure on the right). He is carrying 3 kg of apples in a thin plastic bag.

(a) The newtonmeter indicates 29.5 N. He can therefore conclude correctly that:

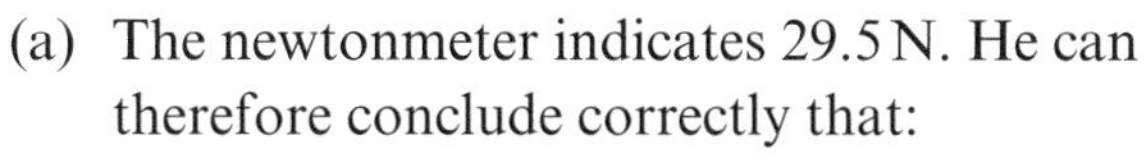

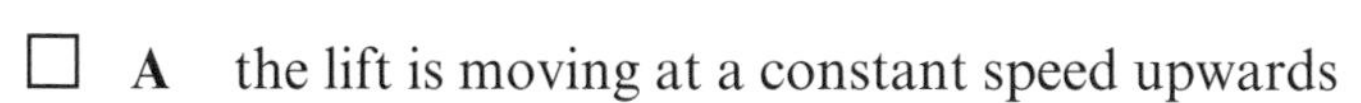

☐ **A** the lift is moving at a constant speed upwards

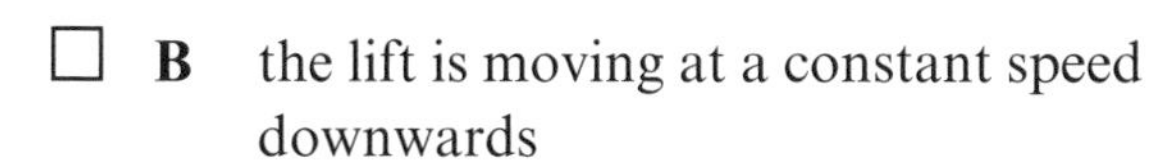

☐ **B** the lift is moving at a constant speed downwards

☐ **C** the lift is moving at a constant speed upwards or downwards

☐ **D** the lift is either accelerating or decelerating.

(1 mark)

(b) The motion of the lift changes and the reading on the newtonmeter is now 25 N (see figure on the right). Isaac should conclude that:

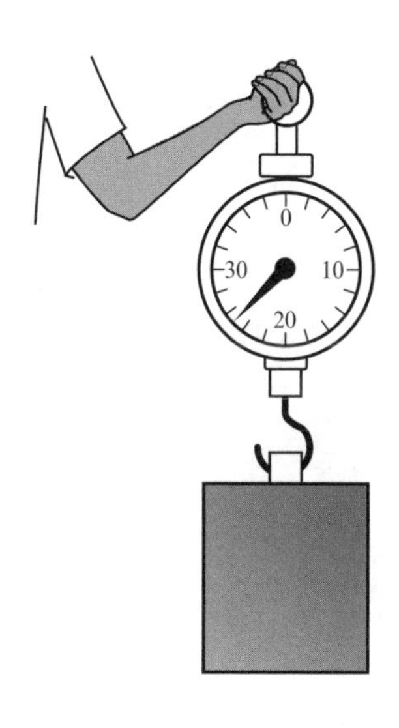

☐ **A** the lift is now moving at a lower constant speed

☐ **B** the lift is now moving at constant speed downwards

☐ **C** the lift is accelerating upwards

☐ **D** the lift is accelerating downwards.

(1 mark)

Guided

(c) Isaac and the apples have a total mass m. The lift now has an **upward** acceleration a. The gravitational field strength is g. The normal contact force between Isaac and the floor of the lift will be:

If the lift is accelerating upwards, the contact force is greater than Isaac's static weight, because the floor is also applying a force that accelerates him upwards.

☐ ~~**A**~~ mg ☐ **C** $m(g - a)$

☐ **B** $m(g + a)$ ☐ **D** $m(a - g)$

(1 mark)

(d) The mass of the lift is M and it now has a **downward** acceleration a. The gravitational field strength is g. The tension in the lift cable is:

☐ **A** Mg ☐ **C** $(M + m)a$

☐ **B** $(M+ m)(g - a)$ ☐ **D** $(M + m)(a + g)$

(1 mark)

Had a go ☐ Nearly there ☐ Nailed it! ☐

Drag

1 The drag force F_D acting on an object of cross-sectional area A moving in a fluid of density ρ at a velocity v is given by: $F_D = \frac{1}{2} C_D \rho A v^2$

C_D is the drag coefficient and is a constant for a particular shape of object.

Guided

(a) Show that the drag coefficient C_D does not have any units.

Rearranging the equation above shows that the drag coefficient has the same units as force/(density × area × speed^2) so ....

....

.... **(2 marks)**

(b) A well-known car manufacturer once claimed that their car had the lowest drag coefficient C_D of any production car. Explain why this is not sufficient information to determine whether their car produced less drag than other cars.

....

....

.... **(2 marks)**

(c) Explain why driving at very high speed has a significantly detrimental effect on the fuel consumption of a vehicle.

....

....

Fuel consumption is very largely dependent on the forces opposing the motion of the car.

.... **(2 marks)**

2 A ball bearing is dropped into a column of oil. The ball initially accelerates but quickly reaches terminal velocity.

Add labelled arrows to the following diagrams to indicate the forces acting when (a) the ball is accelerating and (b) when it has reached terminal velocity. **(4 marks)**

(a) (b)

Remember that the length of an arrow representing a force relates to the size of the force.

(c) Explain why the motion of the ball changes between (a) and (b).

....

....

....

....

....

.... **(3 marks)**

Centre of mass and centre of gravity

1 The centre of mass of an object is best described as:

☐ **A** the point at which the mass of the object can usually be considered to be

☐ **B** the point where the total mass of the object is located

☐ **C** the point inside the object where gravity acts

☐ **D** the geometrical centre of the object. **(1 mark)**

2 The figure represents a plastic traffic cone.

(a) Mark with an 'X' where you think the centre of gravity of the cone is. **(1 mark)**

Guided

(b) What aspects of the cone's design increase its stability in a strong wind?

It has a low centre of mass and ....

....

....

....

(1 mark)

3 The figure shows a painting hanging from a string over a nail. It has been disturbed from its normal horizontal position.

This is a question about moments and centre of gravity.

Explain why the painting hangs at this particular angle. Use the diagram to aid your explanation.

....

....

....

.... **(3 marks)**

Moments, couples and torques

1 A mechanic is trying to loosen a rusty bolt using a spanner as illustrated below. She exerts a force of 50 N as shown.

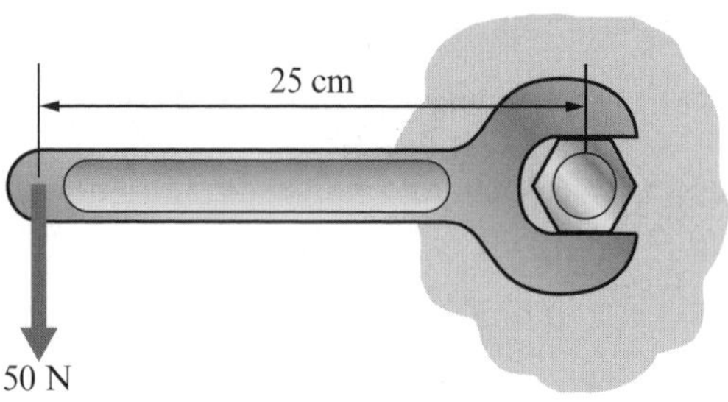

(a) Add an arrow to the diagram to show the force exerted by the bolt on the spanner. **(2 marks)**

(b) State the magnitude of this force.

.... **(1 mark)**

(c) The two forces on the spanner form a couple. Determine the torque of this couple.

....

....

.... **(3 marks)**

2 The figure shows a railway porter moving heavy goods using a 'sack truck'. The total weight of the sack truck and goods is 2.0 kN.

> Consider only vertical forces and horizontal distances. Anything else is irrelevant!

(a) State the principle of moments.

....

....

.... **(2 marks)**

Guided (b) Determine the upward force that the porter must exert to support the load.

Considering moments about the wheel of the sack truck ....

....

.... **(2 marks)**

(c) Determine the force exerted on the ground by the wheels of the sack truck.

....

....

.... **(2 marks)**

Equilibrium

1 A skydiver falling at terminal velocity (see figure right) is not accelerating because:

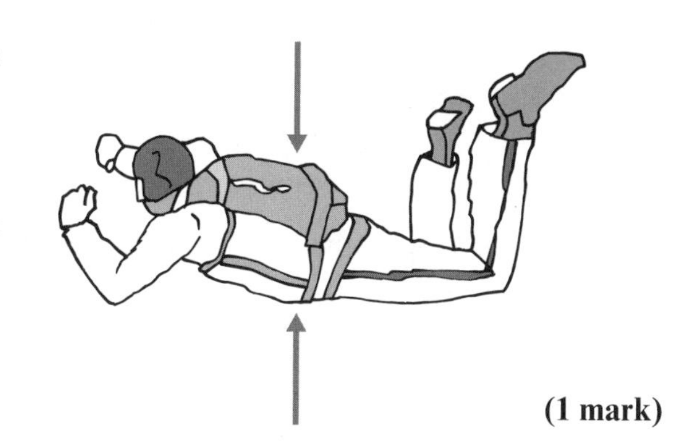

☐ **A** his weight is always constant

☐ **B** the drag force acting on him is constant

☐ **C** his weight is equal in magnitude to the drag force

☐ **D** there are no forces acting on him.

(1 mark)

2 The figures below show forces acting on a spherical object. The arrows show the direction of the forces of the magnitudes specified.

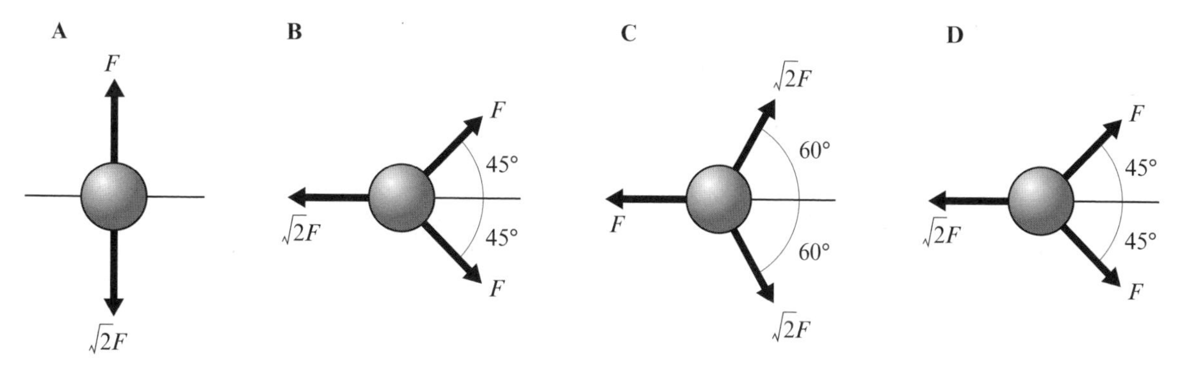

In which of A, B, C, D are the forces shown in equilibrium?

☐ **A** ☐ **B** ☐ **C** ☐ **D**

(1 mark)

3 The figure below shows three forces acting on a metre rule.

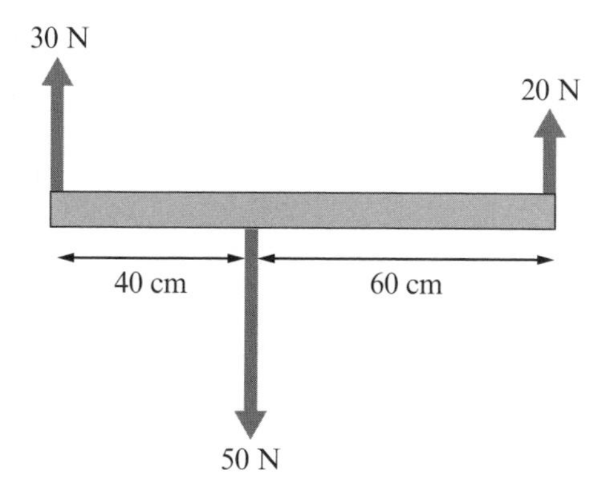

Verify that the metre rule is in both force and rotational equilibrium.

....

....

....

....

.... **(3 marks)**

Had a go ☐ Nearly there ☐ Nailed it! ☐

Density and pressure

1 A student wants to know how much copper wire is left on a part used reel. She measures the diameter of the wire and finds it to be 0.80 mm. The mass of copper on the reel is 0.750 kg. The density of copper is $8.96\,\mathrm{g\,cm^{-3}}$.

(a) Determine the volume of the copper wire on the reel.

....

....

....

(2 marks)

Guided

(b) Determine the cross-sectional area of the copper wire.

The cross-sectional area $= \pi r^2 = \frac{\pi D^2}{4} = \pi \times \frac{(0.80 \times 10^{-3})^2}{4} =$ ....

....

.... **(2 marks)**

(c) Hence determine the length of wire remaining on the reel.

....

....

.... **(2 marks)**

2 A mercury barometer measures atmospheric pressure. Atmospheric pressure acting on the surface of the reservoir is balanced by the pressure due to the column of mercury. Standard atmospheric pressure is 101 kPa. The density of mercury is $13\,600\,\mathrm{kg\,m^{-3}}$.

vacuum

mercury column

h

$P_{atmosphere}$

mercury reservoir

This is a question on hydrostatic pressure: you will need to use the equation $p = h\rho g$.

(a) Determine the height of mercury column that can be supported by one atmosphere.

....

....

.... **(2 marks)**

(b) Why would it be impractical to use a column of water rather than one of mercury?

....

....

.... **(2 marks)**

Upthrust and Archimedes' principle

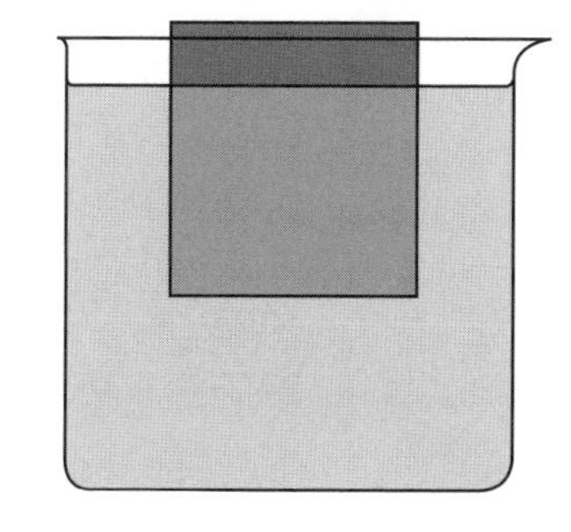

1 A 5.0 cm wooden cube floats in a beaker of water (density = 1000 kg m^{-3}) as shown.

Four-fifths of the block is submerged with only one-fifth above the water's surface.

Guided (a) Determine the pressure at the lower horizontal surface of the block.

Using the equation for hydrostatic pressure:

$p = h\rho g$ .... **(2 marks)**

(b) Determine the upward force on the block. ....

.... **(2 marks)**

(c) State the weight of the block.

.... **(1 mark)**

(d) Determine the density of the wooden block. ....

.... **(2 marks)**

2 A 1.00 kg mass is suspended from a light thread. It is slowly lowered into a beaker of water on a top-pan balance. The reading on the balance increases from 500 g to 625 g.

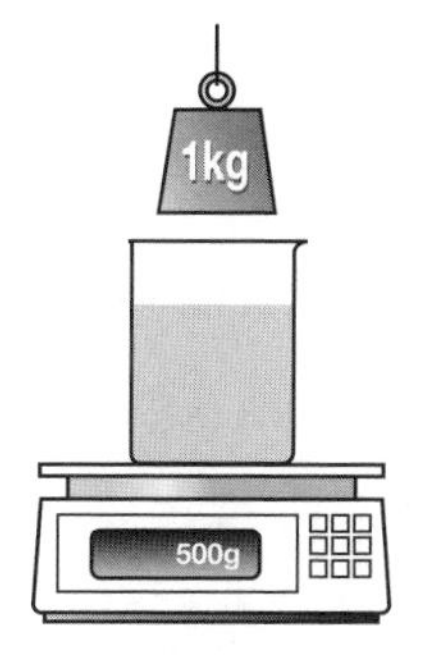

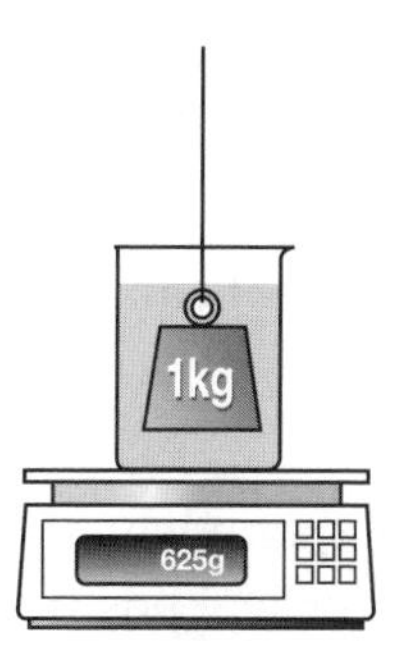

Tension is a force, but the balance indicates a mass, so use $W = mg$ where m is in kg and $g = 9.81\,\text{N kg}^{-1}$.

(a) Determine the tension in the thread before the 1.00 kg mass enters the water.

.... **(2 marks)**

(b) Determine the tension in the string after the 1.00 kg mass is completely submerged but not touching the bottom of the beaker.

.... **(2 marks)**

(c) Account for the change in the reading on the balance as the mass enters the water.

....

.... **(2 marks)**

The mass is now lowered further into the beaker until it touches the bottom and the thread goes slack.

(d) State the reading on the balance. .... **(1 mark)**

Exam skills

1 This question is about using a winch to raise a horizontal security barrier.

The 4.0 m long barrier weighs 200 N. It is raised by a wire rope attached to its mid-point, which is also its centre of gravity. The wire rope makes an angle of 45° to the barrier when the barrier is in the down position.

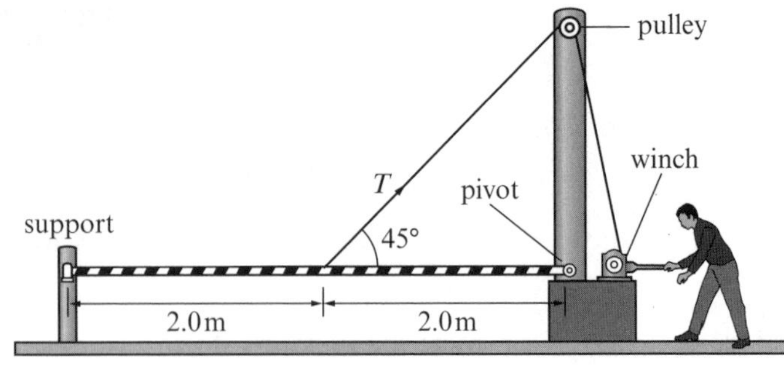

(a) The tension in the wire rope is initially zero and the beam rests on a support. Determine the force exerted on the end of the beam by the support.

....

.... **(2 marks)**

(b) Explain why the resultant force on the barrier due to the pivot and the support must act vertically upwards.

....

.... **(1 mark)**

(c) The winch is turned so that the barrier just leaves contact with the support but is still horizontal. Determine the tension T in the wire rope under these conditions.

....

....

.... **(3 marks)**

(d) Explain why the force at the pivot cannot act vertically upwards.

.... **(1 mark)**

(e) The beam is slowly raised. Explain why the tension in the wire rope decreases as the barrier approaches a vertical position.

....

....

....

.... **(3 marks)**

The figure shows some details of the winch. The barrier is stationary and the tension in the wire rope is 240 N.

(f) Determine the size of the force F.

T = 240 N

6.0 cm

30.0 cm

F

....

....

.... **(2 marks)**

Work done by a force

1 A crate of mass 100 kg is being pushed across a horizontal floor at a steady velocity of 0.5 m s^{-1} with a force of 400 N (see figure).

(a) What is the magnitude of the frictional force acting on the box?

.... **(1 mark)**

(b) Indicate the direction of the frictional force on the diagram. **(1 mark)**

(c) Calculate the work done in pushing the crate 5.0 m along the floor.

....

....

.... **(3 marks)**

(d) What quantity of heat energy is transferred to the surroundings as a result of the work done above?

.... **(1 mark)**

It is decided that pulling the crate with a rope will be easier.

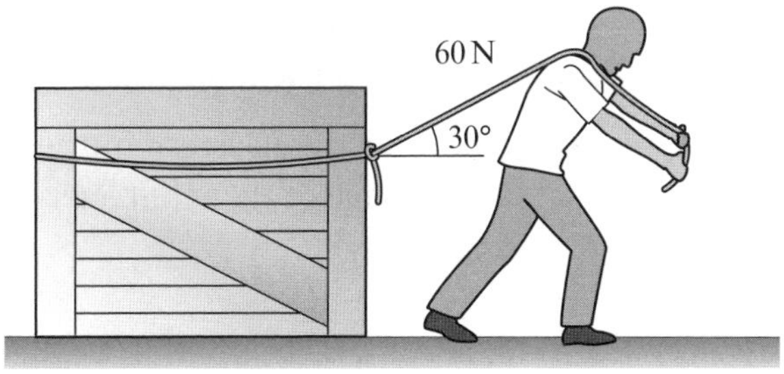

Guided

(e) Why is a larger force now needed to drag the crate along the same floor at the same speed?

Because the pulling force is not horizontal, ...

....

.... **(2 marks)**

(f) How much work is done dragging the crate 5.0 m along the floor?

....

....

.... **(2 marks)**

(g) Compare and account for your answers to (c) and (f) above.

....

....

.... **(2 marks)**

Had a go ☐ Nearly there ☐ Nailed it! ☐

Conservation of energy

1 (a) State the principle of conservation of energy.

....

.... **(2 marks)**

(b) If energy is always conserved, why do we talk about the importance of not 'wasting energy'?

....

....

.... **(2 marks)**

2 Describe the main energy transfers performed by the following devices:

(a) microphone: .... into .... **(2 marks)**

(b) light emitting diode: .... into .... **(2 marks)**

(c) candle: .... into .... and .... **(2 marks)**

3 The figure shows the energy flow associated with a photovoltaic power plant that uses solar panels to convert sunlight into electrical output.

(a) Determine the electrical output as a percentage of the input energy from sunlight.

....

.... **(1 mark)**

(b) The input power from sunlight available for the panels on a particular day is $750\,\text{W m}^{-2}$. What area of panels will be required to produce an electrical output power of 1.8 kW?

....

....

.... **(3 marks)**

(c) Photovoltaic panels are relatively inefficient but are still of great interest to developers. Discuss their advantages and disadvantages.

....

....

....

....

.... **(4 marks)**

Kinetic and gravitational potential energy

1 A force F can do work on an object of mass m such that the only result is that the object accelerates.

Use **equations of motion** ('SUVAT') to show that the energy associated with the motion of an object is given by $E_k = \frac{1}{2}mv^2$.

The v^2 in $\frac{1}{2}mv^2$ in suggests which SUVAT equation you should start with.

....

....

.... **(3 marks)**

2 A car of mass 1800 kg initially travelling at $20\,\mathrm{m\,s^{-1}}$ decelerates uniformly to rest over a distance of 40 m in an emergency stop.

(a) Determine the kinetic energy of the car before it starts to decelerate.

.... **(1 mark)**

Guided

(b) Use your result from (a) to determine the mean braking force during deceleration.

The work done by the brakes is equal to the initial kinetic energy of

the car = ... **(2 marks)**

(c) Describe the main energy transfer that takes place while the car is decelerating.

.... **(2 marks)**

3 A rubber ball is dropped from a height of 1 m onto a horizontal surface and then bounces several times. Its velocity/time graph is reproduced here. Air resistance can be neglected.

Bounce height is determined by the kinetic energy of the ball as it leaves the surface.

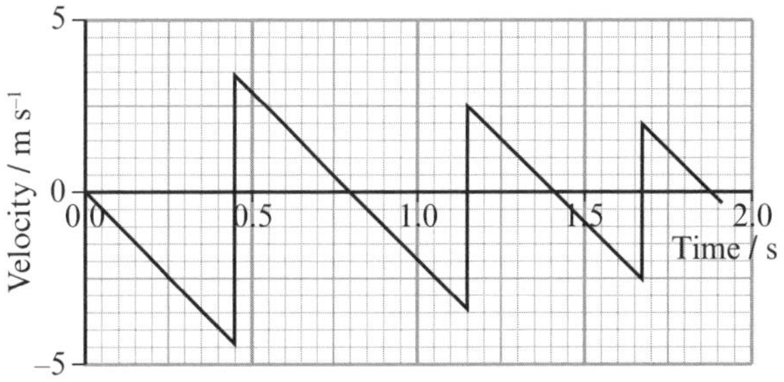

Velocity / m s^{-1}
5
0
−5
0.0
0.5
1.0
1.5
2.0
Time / s

(a) How does the graph show that some kinetic energy is lost as a result of each bounce?

....

.... **(2 marks)**

(b) Why does the bounce height decrease after each bounce?

....

.... **(2 marks)**

(c) Determine the **height** of the first bounce.

.... **(2 marks)**

(d) What percentage of the ball's kinetic energy is lost as a result of the first bounce?

.... **(2 marks)**

Had a go ☐ Nearly there ☐ Nailed it! ☐

Mechanical power and efficiency

1 Which of the following is **not** a unit of power?

☐ **A** $J\,s^{-1}$ ☐ **B** $N\,m\,s^{-1}$ ☐ **C** $kg\,m^2\,s^{-2}$ ☐ **D** $kg\,m^2\,s^{-3}$ **(1 mark)**

Guided

2 If a force F acts on an object travelling at a constant speed v, show that the required power P to maintain this motion is given by $P = Fv$.

Work done is equal to force × distance so the rate at which work is

done (the power) is equal to ....

....

.... **(2 marks)**

3 A particular car requires a mechanical output power of 8.0 kW when driving at 20 m s^{-1} on a level road. At 30 m s^{-1}, it requires an output power of 22.0 kW.

(a) Determine the driving force required at:

(i) 20 m s^{-1}

....

.... **(2 marks)**

(ii) 30 m s^{-1}

....

.... **(1 mark)**

(b) Why does the driving force change as the car accelerates from 20 m s^{-1} to 30 m s^{-1}?

....

.... **(2 marks)**

(c) If 1.0 kg of fuel provides 40 MJ of input energy and the overall efficiency of the car is 20%, how much fuel would be used when driving a journey of 40 km at 20 km/s?

....

....

Remember to take into account the 20% efficiency in the calculation.

....

....

.... **(4 marks)**

(d) If the mass of the car is 1600 kg, what additional output power will be required for it to continue driving at 20 m s^{-1} up a 5° gradient?

....

....

.... **(3 marks)**

Exam skills

1 This question is about driving piles into the ground to reinforce the foundations of a building. The simplest 'drop hammer' method involves lifting a heavy hammer and then simply allowing it to free-fall and strike the pile cap as shown.

hammer
guides
pile
pile cap

The mass of the hammer is 2500 kg. It is raised to a height of 4.0 m above the top of the pile before it is dropped and falls freely. Each time the hammer strikes the pile cap, the pile is driven 0.12 m further into the ground.

(a) Determine the gravitational potential energy gained by the hammer each time it is lifted above the pile.

....

.... **(2 marks)**

(b) State the kinetic energy of the hammer immediately before it strikes the top of the pile.

.... **(1 mark)**

When the hammer strikes the top of the pile, 80% of its kinetic energy is available to do work in driving the pile into the ground.

(c) Determine the average force between the pile and the ground as it is driven into the ground.

....

....

.... **(3 marks)**

(d) To what form of energy is the kinetic energy of the hammer transferred once the hammer has stopped moving?

.... **(1 mark)**

A more sophisticated pile-driver based on the diesel engine principle can deliver 40 blows per minute to the top of the pile with an energy of 120 kJ per blow.

(e) Determine the average power output of the diesel-driven pile-driver.

....

.... **(2 marks)**

The pile-driver uses 10 kg of diesel fuel per hour when operating. 1 kg fuel provides 48 MJ of input energy.

(f) Determine the overall efficiency of the diesel-driven pile-driver.

....

....

.... **(2 marks)**

Elastic and plastic deformation

1 Explain the difference between elastic and plastic deformation of a material.

....

....

.... **(3 marks)**

2 The graph shows the behaviour of a spring under increasing loads.

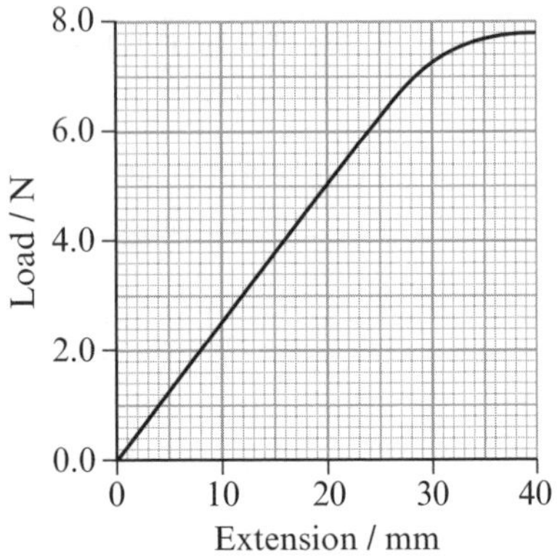

(a) Determine the load and extension when the spring reaches its limit of proportionality.

....

.... **(2 marks)**

Load / N
8.0
6.0
4.0
2.0
0.0
0
10
20
30
40
Extension / mm

Guided

(b) Determine the force constant of the spring.

The force constant k of a spring is defined by the equation $F = kx$, so $k = \frac{F}{x}$ or the slope of the straight part of the graph.

.... **(2 marks)**

(c) Determine the elastic potential energy stored in the spring when it is extended by 20 mm.

....

.... **(2 marks)**

3 The graph shows the behaviour of a ductile material under increasing tensile force.

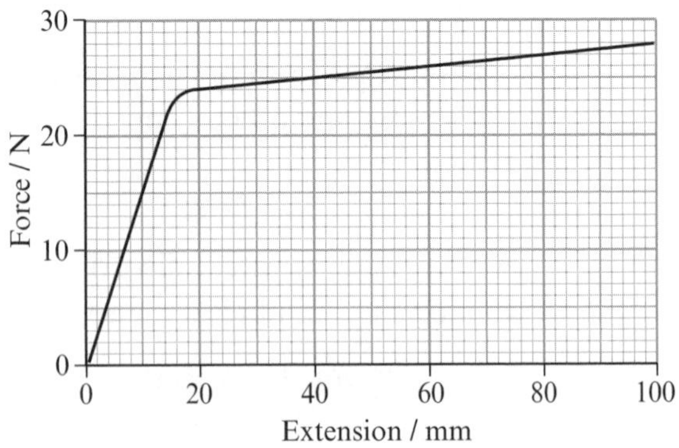

(a) At what value of tension does the material reach its limit of proportionality?

....

(1 mark)

Force / N
30
20
10
0
0
20
40
60
80
100
Extension / mm

(b) The material deforms plastically for extensions above 20 mm. How much work is done in increasing the extension from 20 mm to 100 mm?

....

.... **(2 marks)**

(c) While extending from 20 mm to 100 mm, very little of the work done goes into additional stored elastic potential energy. What happens to the remaining energy?

....

.... **(2 marks)**

Stretching things

1 An experiment is to be done to determine the elastic behaviour of a spring and find its force constant. The apparatus used is similar to the one shown.

Describe how you would conduct the experiment in order to obtain an accurate and reliable result. You should state what measurements need to be taken and how these results are obtained. You may add to the diagram where necessary.

Look for hints in the question if you are asked to describe an experiment, for example key words like 'accurate' and 'reliable'. State what you will measure, how you will measure it and what you will do with the data.

....

....

....

....

....

....

....

....

....

....

....

.... **(6 marks)**

2 In an experiment to investigate the stretching of a copper wire under load, it is advisable to use thin wire and as long a length as possible.

(a) Explain how these two factors contribute to the overall percentage uncertainty of the experiment.

....

....

....

.... **(3 marks)**

(b) We also want to find the cross-sectional area of the wire. Why might using a very thin wire have a detrimental effect on the uncertainty of this part of the experiment?

....

....

.... **(2 marks)**

Had a go ☐ Nearly there ☐ Nailed it! ☐

Force–extension graphs

1 The graph shows the behaviour of a rubber band under increasing and decreasing loads.

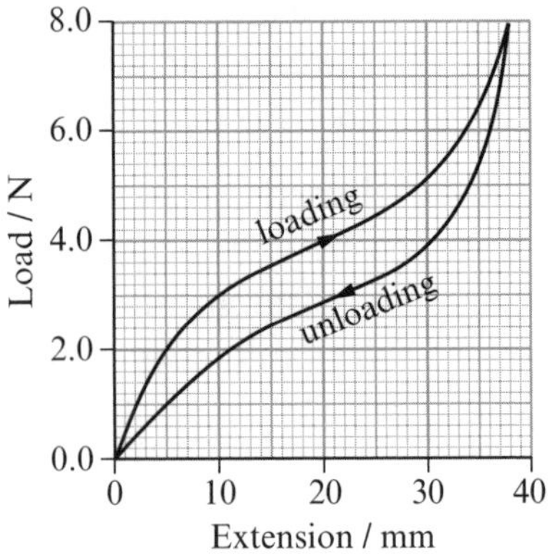

Load / N: 0.0, 2.0, 4.0, 6.0, 8.0
Extension / mm: 0, 10, 20, 30, 40
loading
unloading

(a) How does this graph show that the rubber band does not obey Hooke's law?

....

....

....

(2 marks)

(b) How does the graph show that the rubber is behaving elastically?

....

....

.... **(2 marks)**

(c) Describe how the stiffness of the rubber changes as the load increases from zero to 8.0 N.

'Stiffness' means resistance to stretching in this context.

....

....

.... **(2 marks)**

(d) Use the graph to estimate the work done in stretching the rubber band to an extension of 20 mm.

'Estimate' means make a rough estimate of the area under the line on the graph. You do not need to count every tiny square.

....

....

.... **(2 marks)**

(e) How does the graph show that more work is done in stretching the rubber band than is done by the rubber band when it is unstretched back to its original state?

....

....

.... **(2 marks)**

(f) A car's tyres can affect its overall efficiency. Use your answer to (e) above to help explain why the tyres contribute to car's energy losses.

....

....

.... **(2 marks)**

Stress and strain

1 Two rods, A and B, made of the same material, are put under tension by equal and opposite forces, F. B has twice the length of A and twice the diameter of A (see figure below).

F ◄— A — B —► F

(a) Which of the following statements is true?

As the rods are in series, the same force acts on each, so the stress will only depend on the cross-sectional area.

☐ **A** The tensile stress in each rod is the same.

☐ **B** The tensile stress in A is twice the tensile stress in B.

☐ **C** The tensile stress in A is four times the tensile stress in B.

☐ **D** The tensile stress in A is eight times the tensile stress in B. **(1 mark)**

(b) Which of the following statements is true?

☐ **A** The resultant strain in A is the same as the resultant strain in B.

☐ **B** The resultant strain in A is twice the resultant strain in B.

☐ **C** The resultant strain in A is four times the resultant strain in B.

☐ **D** The resultant strain in A is eight times the resultant strain in B. **(1 mark)**

2 A beam weighing 2000 N is being supported by two cables. These cables are in turn connected to a single vertical cable (see figure).

T 60° 60° T

(a) State the tension in the vertical cable.

....

(1 mark)

Guided

(b) Determine the tension T in each of the other two cables.

Equating the vertical forces gives $2T\cos 60° = 2000$ and so ....

....

.... **(2 marks)**

(c) A cable will fail if the stress in it exceeds 800 MPa. If the area of cross-section of each cable is 20 mm^2, show that the beam can be lifted without risk of any of the cables failing.

....

.... **(3 marks)**

(d) What is the maximum weight of beam that can be lifted in this manner if the maximum allowable tensile stress in any cable is 50% of the stress at failure?

....

.... **(2 marks)**

Had a go ☐ Nearly there ☐ Nailed it! ☐

Stress–strain graphs and the Young modulus

1 The Young modulus of aluminium is 70 GPa while that of copper is 130 GPa. From this information we can conclude that:

☐ **A** copper is stronger than aluminium

☐ **B** copper is more ductile than aluminium

☐ **C** copper is stiffer than aluminium

☐ **D** copper is denser than aluminium. **(1 mark)**

2 This question is about **stress** and **strain** in relation to deforming a material.

(a) Define **tensile stress** and explain why its units are the same as those of pressure.

....

.... **(2 marks)**

(b) Define **tensile strain** and explain why strain does not have any units.

....

.... **(2 marks)**

(c) Hence explain why the Young modulus has the same units as stress.

....

.... **(2 marks)**

3 The figure is a stress–strain graph for an aluminium alloy.

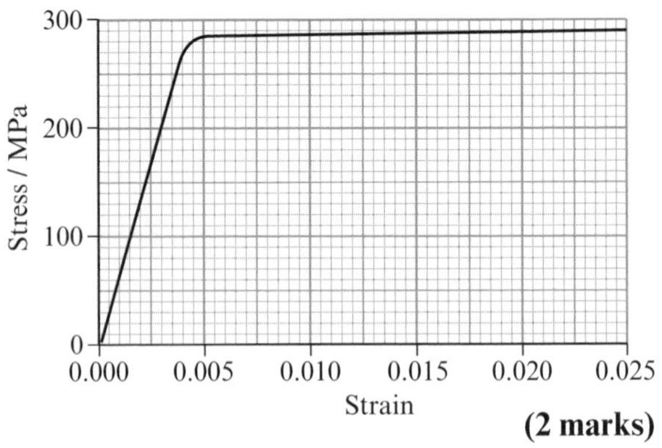

(a) The alloy's yield stress is 280 MPa. Mark the yield point on the graph and explainthe significance of the term yield stress.

....

.... **(2 marks)**

(b) Use the graph to determine the Young modulus of the aluminium alloy.

....

.... **(2 marks)**

(c) In a test, a 100 mm long sample of the alloy is subjected to a stress of 290 MPa. The stress is then removed. Use the graph to determine the length of the sample after the test.

....

.... **(3 marks)**

Measuring the Young modulus

1 The apparatus shown is to be used to measure the Young modulus of copper.

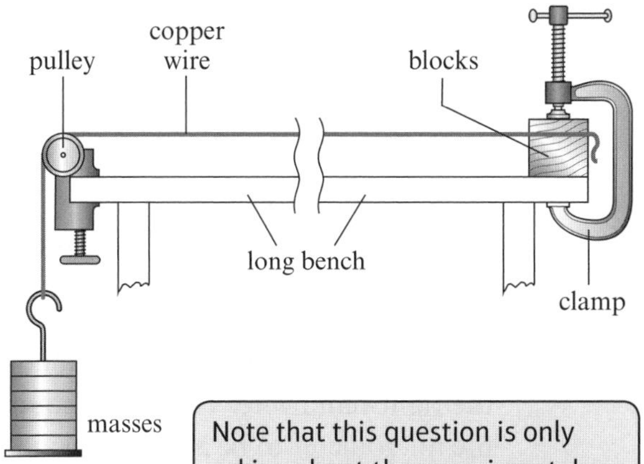

(a) Describe the experimental procedure, including any additional equipment that might be needed, and stating what measurements need to be taken.

Note that this question is only asking about the experimental procedure and not how you deal with the experimental data once you have it. Don't do unnecessary work that earns no marks.

.... **(5 marks)**

(b) Describe how the above data should be processed to obtain an accurate value for the Young modulus of copper.

Now you can say what you do with the data. That usually involves plotting a graph to get the gradient.

.... **(3 marks)**

2 The Young modulus of copper is 120 GPa. A 3.0 m length of copper wire of diameter 0.50 mm is subjected to a tensile force of 10 N.

(a) Determine the tensile stress in the wire. ....

.... **(2 marks)**

(b) Determine the resultant extension of the wire. ....

.... **(2 marks)**

(c) The yield stress of copper is 70 MPa. Determine the maximum tensile force that can be applied to the wire before it starts to deform plastically.

.... **(2 marks)**

(d) The tensile strength of copper is 220 MPa. Why is it not possible to predict the maximum extension of the wire immediately before it fails?

.... **(2 marks)**

Had a go ☐ Nearly there ☐ Nailed it! ☐

Exam skills

1 This question is about measuring the Young modulus of steel using Searle's method.

The apparatus consists of two long, parallel, identical steel wires suspended from a rigid bracket at the upper end and attached to a measuring device at the lower end (see figure). At the start of the experiment, the levelling bubble is centred using the micrometer. The variable mass can then be increased, which extends the right-hand wire. This then requires the micrometer to be adjusted again by exactly the extension of the wire in order to level the bubble. This enables the extension of the wire to be measured accurately as the load is increased.

Load / kg	extension / mm
0.00	0.00
1.00	0.47
2.00	0.99
3.00	1.51
4.00	1.97
5.00	2.46

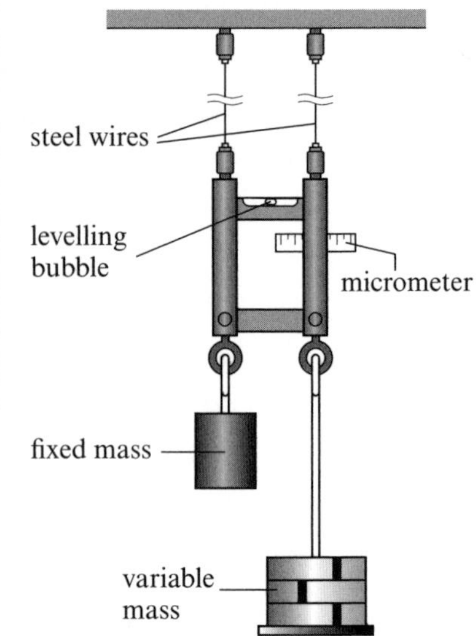

(a) Explain how using two parallel wires can reduce the effect of changes on the accuracy of the experiment.

....

....

.... **(2 marks)**

(b) Plot a graph of mass against extension using the data provided in the table above. **(4 marks)**

(c) Determine the gradient of the graph.

....

.... **(2 marks)**

The wire has diameter of 0.60 mm and a length of 3.00 m.

(d) Determine the cross-sectional area of the wire in m^2.

....

.... **(2 marks)**

(e) Show that the Young modulus of steel is about 210 GPa.

....

....

....

.... **(2 marks)**

Newton's laws of motion

1 According to Newton's first law of motion, in the absence of a resultant force, an object:

☐ **A** cannot ever move

☐ **B** will accelerate uniformly

☐ **C** will decelerate until its velocity is zero

☐ **D** will move with a constant velocity. **(1 mark)**

2 A ball can bounce upwards off a hard horizontal surface because shortly after impact:

☐ **A** the upward force on the ball is greater than the downward force on the surface

☐ **B** the downward force on the surface is greater than the upward force on the ball

☐ **C** the upward force on the ball is greater than the downward force on the ball

☐ **D** the forces acting on the ball and the floor are equal and opposite. **(1 mark)**

3 A man is standing on a set of bathroom scales in a lift that is accelerating vertically downwards. Which of the following statements must be true?

☐ **A** The reading on the scales is equal to the weight of the man.

☐ **B** The upward force of scales on the man is equal to the weight of the man.

☐ **C** The reading on the scales is equal to the upward force of the scales on the man.

☐ **D** The sum of all the forces acting on the man is zero. **(1 mark)**

4 A satellite moves in a circular orbit around the Earth at a constant speed. Which of the following statements must be true?

☐ **A** There must be a resultant force acting on the satellite.

☐ **B** There are no forces acting on the satellite.

☐ **C** The resultant force on the satellite is zero.

☐ **D** The satellite is beyond the pull of gravity.

(1 mark)

Had a go ☐ Nearly there ☐ Nailed it! ☐

Linear momentum

1 Which of the following has the greatest linear momentum?

☐ **A** A 45 kg boy running at $9\,m\,s^{-1}$

☐ **B** A 45 kg cheetah running at $90\,km\,h^{-1}$

☐ **C** A 4.5 g bullet travelling at $900\,m\,s^{-1}$

☐ **D** A 0.045 kg arrow travelling at $90\,m\,s^{-1}$ **(1 mark)**

2 Which of the following is **not** a possible unit of momentum?

☐ **A** Pa s ☐ **B** $J\,m^{-1}\,s$ ☐ **C** N s ☐ **D** $kg\,m\,s^{-1}$ **(1 mark)**

3 A horizontal conveyor belt in a steel factory runs at $4\,m\,s^{-1}$ and delivers 500 kg of iron in every 10 seconds into a large vessel. The force exerted by the conveyor belt on the iron ore is:

☐ **A** 20 N ☐ **B** 200 N ☐ **C** 2000 N ☐ **D** 20 000 N **(1 mark)**

4 A 1500 kg car travelling at $12\,m\,s^{-1}$ is in collision with a concrete bridge support. It can be assumed that the car decelerates uniformly over a distance of 0.60 m before stopping (see figure below).

Remember that when an object decelerates uniformly, the initial speed is twice the average speed.

Guided (a) Determine the duration of the impact.

time = $\frac{\text{distance}}{\text{average speed}} = \frac{0.60}{6}$ = ...

.... **(2 marks)**

(b) Determine the initial momentum of the car.

....

.... **(2 marks)**

Guided (c) Hence find the size of the average force exerted on the car by the wall.

force = $\frac{\text{momentum change}}{\text{time}}$ = ...

.... **(2 marks)**

(d) Suggest two reasons why the force on the driver will be considerably lower than the value in (c) above.

....

....

.... **(2 marks)**

Impulse

1 A 600 g hammer moving vertically downwards at $40\,m\,s^{-1}$ strikes a nail. The hammer stops moving without rebounding after 0.5 ms. It can be assumed that the force exerted by the hammer remains constant during the impact.

(a) Determine the momentum of the hammer just before it strikes the nail.

.... **(2 marks)**

(b) Define the impulse of a force. ....

.... **(1 mark)**

(c) State the impulse of the force of the hammer striking the nail.

.... **(1 mark)**

(d) Determine the force exerted by the hammer on the nail.

.... **(2 marks)**

(e) Determine how far the nail penetrates the wood as a result of the impact.

....

.... **(2 marks)**

2 The graph shows the total force that a sprinter exerts on the starting blocks at the start of a 100 m race. The starting gun is fired at time = 0 s. His body mass is 80 kg.

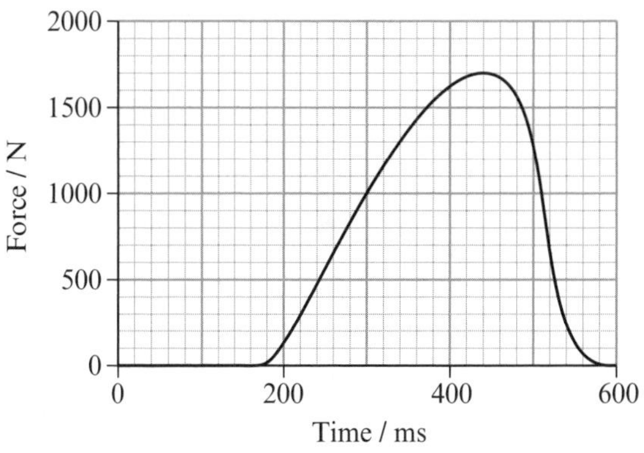

(a) State the reaction time of the sprinter.

.... **(1 mark)**

(b) For how long is the sprinter in contact with and pushing on the starting blocks?

.... **(1 mark)**

(c) Determine the peak acceleration of the sprinter.

....

.... **(2 marks)**

Guided

(d) Estimate the momentum change of the sprinter while he is in contact with the blocks and hence determine the speed of the sprinter as he leaves the blocks.

The area under the force–time graph gives the change in momentum of the sprinter. Each small square is equal to 100 N × 0.020 s = 2.00 N s, so the total change in momentum is ... .

....

.... **(3 marks)**

Conservation of linear momentum – collisions in one dimension

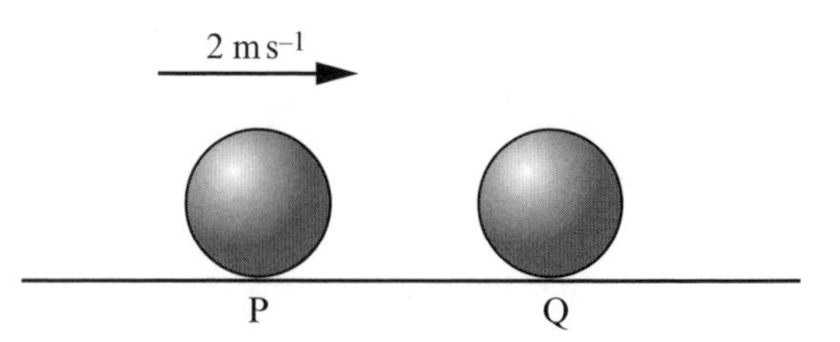

1 Two balls, P and Q, have equal mass. Ball P, moving at 2 m s^{-1} to the right, collides with ball Y, which is initially stationary.

Which of A, B, C and D is a possible outcome after they have collided?

	P	Q
☐ **A**	2 m s^{-1} to the right	stationary
☐ **B**	2 m s^{-1} to the left	stationary
☐ **C**	stationary	2 m s^{-1} to the right
☐ **D**	1 m s^{-1} to the left	1 m s^{-1} to the right

(1 mark)

2 Two balls of mass $2m$ and m are held either side of a horizontal compressed spring. When both balls are released simultaneously, the right-hand ball has an initial velocity of 1 m s^{-1} to the right.

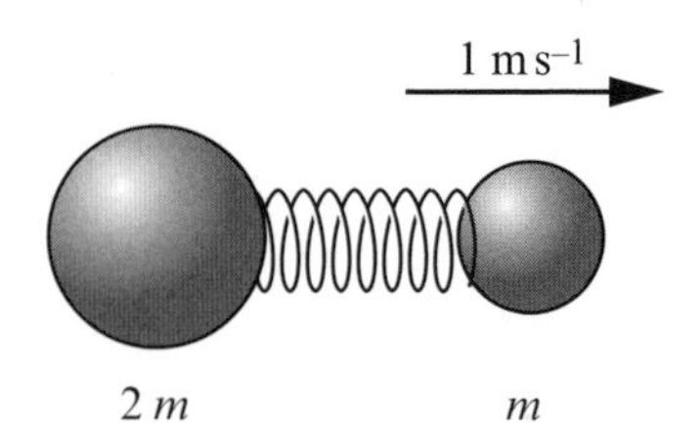

The **initial** velocity of the left-hand ball is:

☐ **A** 1 m s^{-1} to the right

☐ **B** 1 m s^{-1} to the left

☐ **C** 2 m s^{-1} to the left

☐ **D** 0.5 m s^{-1} to the left.

(1 mark)

3 The collision between a golf ball and a club involves a club head of mass 200 g striking a ball of mass 45 g at 31 m s^{-1}. The ball leaves the face of the club at 60 m s^{-1}.

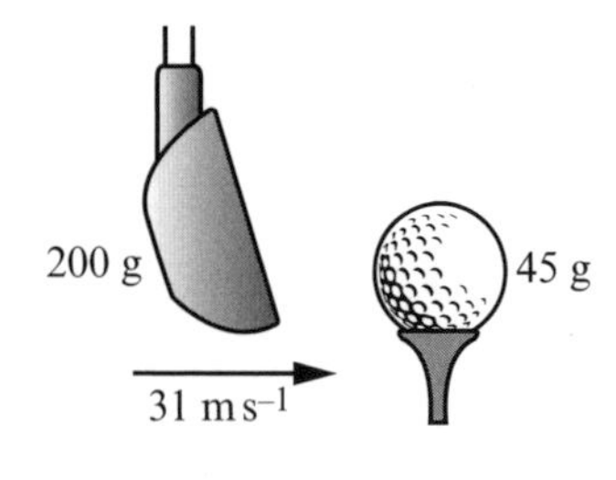

Just apply conservation of momentum before and after the collision, but remember that momentum and velocity are vectors and have direction as well as magnitude.

Guided

(a) Show that the velocity of the club head immediately after the ball leaves contact with it is about 18 m s^{-1}.

Conservation of momentum gives total momentum before impact = total momentum after impact = $0.200 \times 31 = 0.200 \times v_{club} + 0.045 \times 60$

....

(2 marks)

(b) If the club head and ball are in contact for 0.5 ms, determine the force exerted on the ball by the club head.

....

....

....

(2 marks)

Collisions in two dimensions

1 An ice hockey puck of mass 180 g slides with negligible friction while colliding with and rebounding off the smooth side wall of the rink. The collision can be assumed to be perfectly elastic.

(a) Explain why the collision with the wall does not cause any change in the component of the momentum of the puck in a direction parallel to the wall.

....

....

....

....

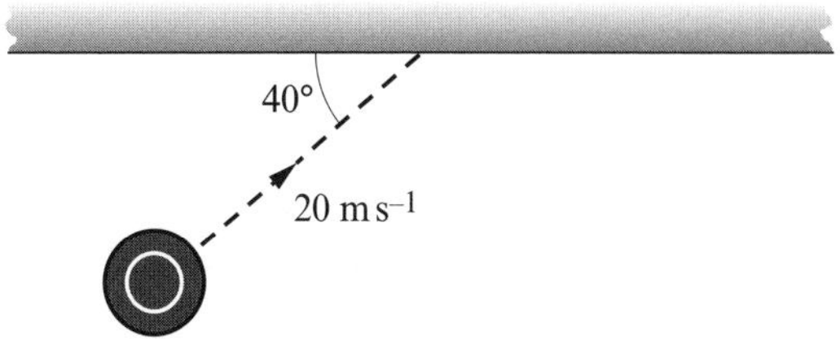

40°

20 m s^{-1}

(2 marks)

Guided

(b) Determine the component of the puck's momentum perpendicular to the wall before the collision.

The perpendicular component will be $mv \sin \theta = 0.180 \times 20 \times \sin 40° =$

....

.... **(2 marks)**

(c) State the component of momentum of the puck perpendicular to the wall after the collision.

....

.... **(1 mark)**

(d) Hence determine the change in momentum of the puck as a result of colliding with the wall of the rink.

....

....

.... **(2 mark)**

2 A subatomic particle X decays spontaneously into particle Y and a less massive particle Z. Which of the paths A, B, C or D in the diagram below could be the track of particle Z?

☐ **A**

☐ **B**

☐ **C**

☐ **D**

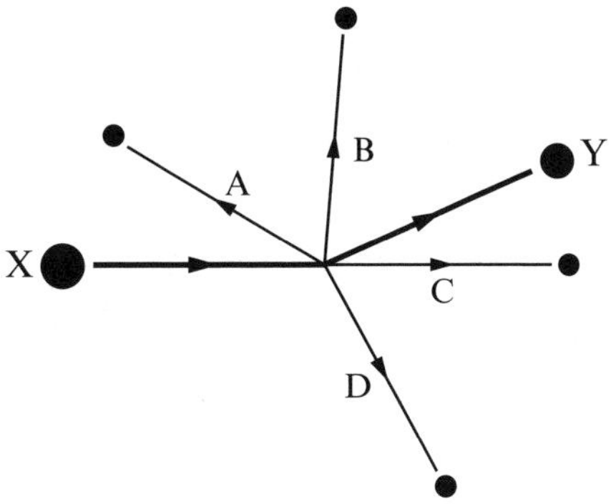

X Y A B C D

In order to conserve momentum both in the direction of X and in a perpendicular direction, there can be only one possible answer.

(1 mark)

Elastic and inelastic collisions

1 Which of the following statements is true concerning a collision between two objects when no external forces act?

☐ **A** The total momentum of both objects is constant only in perfectly elastic collisions.

☐ **B** The total kinetic energy of both objects is constant only in perfectly elastic collisions.

☐ **C** The total momentum of both objects is not constant in inelastic collisions.

☐ **D** The kinetic energy of each object is always constant in perfectly elastic collisions.

(1 mark)

2 A car of mass 1500 kg travelling at $10\,\mathrm{m\,s^{-1}}$ collides with a stationary van of mass 2500 kg. At the moment of impact, both vehicles can be considered to move together as one object.

Guided

(a) Determine the velocity at which the car and van move at the moment of impact.

velocity = $\dfrac{\text{momentum before collision}}{\text{total mass after the collision}}$ = ...

....

.... **(2 marks)**

(b) Determine the kinetic energy of the car before the collision.

....

.... **(1 mark)**

(c) Determine the combined kinetic energy of the car and van at the moment of impact.

....

.... **(1 mark)**

(d) Account for any difference between your answers for (b) and (c) above.

....

....

.... **(2 marks)**

3 For a mass travelling at speed the momentum is given by $p = mv$ and the kinetic energy is given by $E_k = \frac{1}{2}mv^2$. Which of the following shows the correct relationship between p and E_k?

☐ **A** $p = 2m\sqrt{E_k}$

☐ **B** $E_k = \dfrac{p^2}{2m}$

☐ **C** $p = \sqrt{2}mE_k$

☐ **D** $E_k = \dfrac{2p^2}{m}$

(1 mark)

Electric charge and current

1 Which of the following is an SI base unit?

☐ **A** coulomb ☐ **B** volt ☐ **C** ampere ☐ **D** joule **(1 mark)**

2 A simple series circuit consists of a cell and two lamps.

Which of the following statements is true?

☐ **A** The current through lamp A is twice the current through lamp B.

☐ **B** The current through the cell is the sum of the current in lamps A and B.

☐ **C** The same amount of charge flows through lamps A and B in any given time.

☐ **D** The charge flowing through the battery is used up by lamp A and lamp B. **(1 mark)**

A

B

Guided

3 A vacuum tube diode consists of a heated filament cathode and an anode. A conventional current can pass from the anode to the cathode but not in the reverse direction. Given that the electronic charge is 1.60×10^{-19} C, determine the number of electrons per second passing through the diode when the current between anode and cathode is 4 mA.

anode

cathode

The number of electrons per second = charge per second (i.e. current)/charge per electron ....

.... **(2 marks)**

4 A small torch is powered by a battery consisting of two 1.5 V cells in series. Each cell has a capacity of 1800 mA h so they can each provide a current of 1800 mA for one hour, or a different current for a different time, such that the total charge that flows through the battery is constant from fully charged to fully discharged.

(a) Determine the total charge that flows through the battery from fully charged to fully discharged.

....

.... **(2 marks)**

(b) The torch will run for 4 hours if left switched on. Determine the average current that the battery needs to provide over that time.

.... **(1 mark)**

(c) The torch is replaced by a new model that uses the same battery but has a light-emitting diode (LED) rather than a filament bulb. The LED requires a current of 200 mA. For how long will the new torch run if left switched on?

.... **(1 mark)**

Had a go ☐ Nearly there ☐ Nailed it! ☐

Charge flow in conductors

1 Explain why metals are good conductors.

....

....

.... **(2 marks)**

2 Explain how the resistance of a metallic conductor changes as its temperature increases.

....

....

....

....

.... **(3 marks)**

3 Explain why glass is an insulator at room temperature.

....

....

....

.... **(2 marks)**

4 Two electrodes are inserted into an electrolyte such as a dilute acid and a current *I* flows as in the figure.

I

(a) Label the anode and cathode. **(1 mark)**

(b) The current in the wires results from a flow of:

☐ **A** negative ions ☐ **C** positive ions

☐ **B** protons ☐ **D** electrons. **(1 mark)**

(c) The current in the electrolyte results from a flow of:

☐ **A** anions ☐ **C** ions

☐ **B** protons ☐ **D** electrons. **(1 mark)**

(d) Cations are:

☐ **A** positive ions ☐ **C** negative ions

☐ **B** protons ☐ **D** electrons. **(1 mark)**

Kirchhoff's first law

1 Four currents enter or leave a node in a circuit, as shown below.

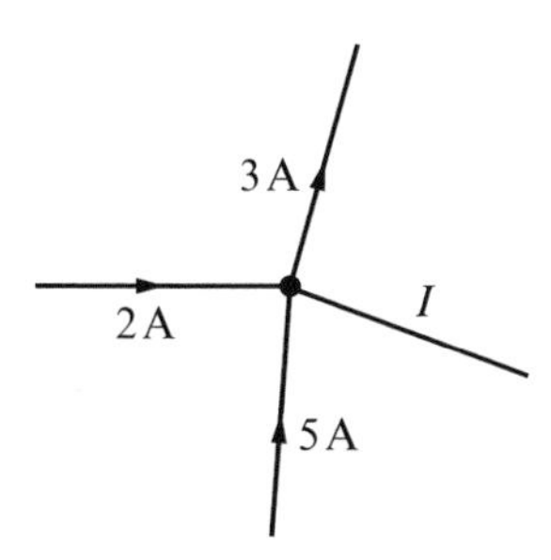

Determine the magnitude of the unknown current, I, and clearly indicate its direction on the above diagram.

.... **(2 marks)**

2 This circuit controls model traffic lights. Different combinations of light-emitting diodes require different switches to be operated, leading to different currents in different parts of the circuit.

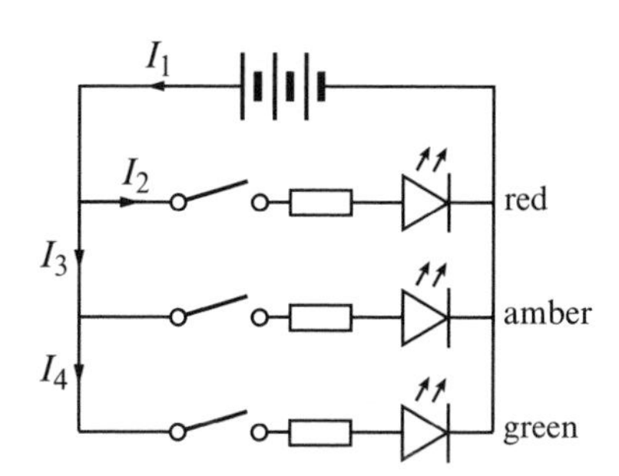

Complete the following table:

LED combination	I_1 / mA	I_2 / mA	I_3 / mA	I_4 / mA
Red		20		
Red + amber	45			
Green			18	
Amber				

(4 marks)

3 Consider the circuit below:

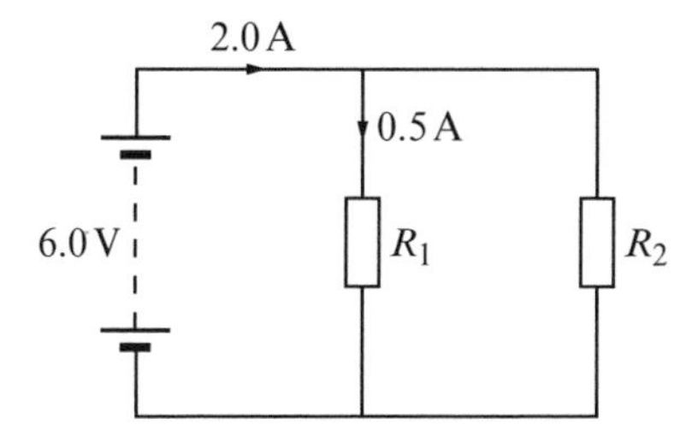

(a) Determine the resistance of R_1.

....

.... **(1 mark)**

(b) Determine the resistance of R_2.

....

....

.... **(2 marks)**

Charge carriers and current

1 A current of 100 A flows through a copper bar as shown below.

(Density of copper is 8.96 g cm^{-3}; molar mass of copper is 63.5 g mol^{-1})

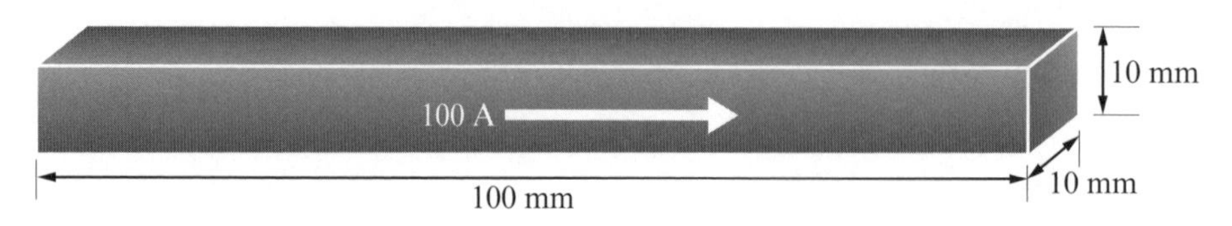

(a) Calculate the mass of the copper bar.

....

.... **(1 mark)**

Guided

(b) Determine the number of moles of copper in the bar.

$$\text{number of moles} = \frac{\text{mass in g}}{\text{molar mass in g mol}^{-1}}$$

....

.... **(1 mark)**

(c) Determine the number of atoms of copper in the bar.

> 1 mole contains 6.02×10^{23} (Avogadro's number) particles, in this case atoms of copper.

....

.... **(1 mark)**

(d) Assuming that one atom of copper provides one conduction electron, state the number of free charge carriers in the bar.

.... **(1 mark)**

(e) How many free charge carriers are present in 1 m^3 of copper?

....

.... **(1 mark)**

(f) What is the drift velocity of charge carriers associated with the current of 100 A flowing through the copper bar?

....

....

....

.... **(2 marks)**

Electromotive force and potential difference

1 Explain what is meant by the **electromotive force** of a **battery**.

....

....

.... **(2 marks)**

2 Which of the following derived units is equivalent to the volt?

A volt is a joule per coulomb. A joule is a newton metre or a kilogram metre^2 second^{-2} and a coulomb is an amp second.

☐ **A** $kg\,m^2\,s^{-3}$ ☐ **B** $A\,s$ ☐ **C** $kg\,m^2\,s^{-1}\,A$ ☐ **D** $kg\,m^2\,s^{-3}\,A^{-1}$ **(1 mark)**

3 A 9 V battery is connected across a 6 V lamp with a resistor in series with it such that the lamp lights normally (see figure below). The current flowing around the circuit is 0.050 A.

9 V

0.050 A

6 V

(a) Determine the charge that flows around the circuit in 100 s.

....

.... **(2 marks)**

(b) Calculate the electrical energy produced by the battery in 100 s.

....

.... **(2 marks)**

(c) Determine the electrical energy converted into heat and light by the lamp in 100 s.

....

.... **(1 mark)**

(d) Explain **in terms of energy** why the potential difference across the resistor must be 3 V.

....

....

....

.... **(3 marks)**

Had a go ☐ Nearly there ☐ Nailed it! ☐

Resistance and Ohm's law

1 Define the quantity **resistance**. ....

.... **(1 mark)**

2 The circuit shown below can be used to measure the electrical characteristics of an electrical component and can also be used to determine its resistance. In this instance, the behaviour of copper sulfate solution is being investigated:

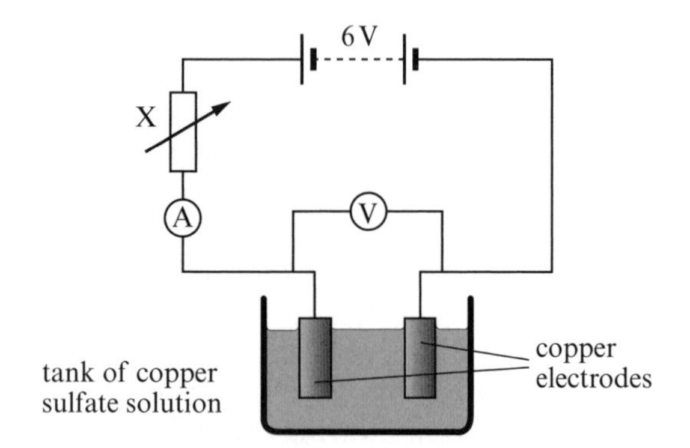

p.d. / V	Current / mA
0.00	0
1.13	74
2.21	147
3.30	223
4.38	301
5.51	375

(a) Name component X and describe its function in this application.

....

.... **(2 marks)**

(b) Name two control variables that must be kept constant during the experiment.

.... **(2 marks)**

(c) The data obtained for the current through the copper sulfate solution and the potential difference between the electrodes is listed to the right of the figure above. Plot a graph of current against potential difference using these axes. **(4 marks)**

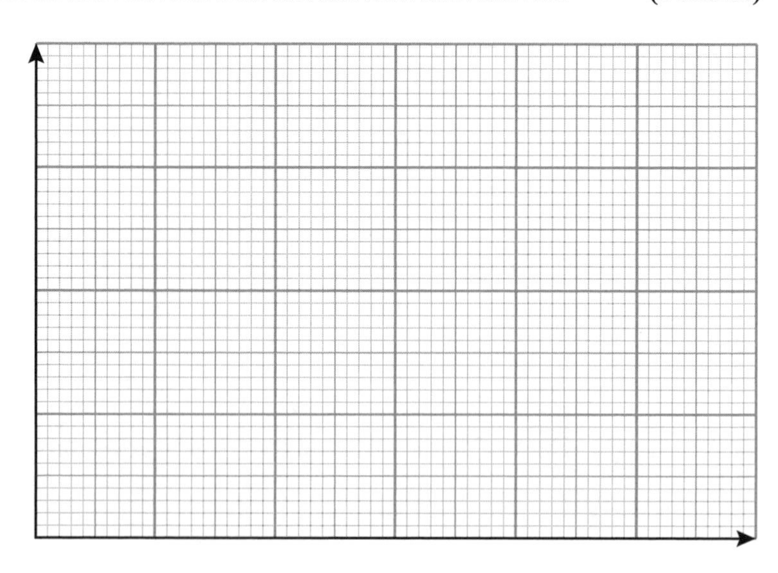

Guided

(d) Explain why it is appropriate to conclude that the copper sulfate solution with copper electrodes is obeying Ohm's law.

The graph is a straight line through the origin so ....

.... **(2 marks)**

(e) Use the graph to determine the resistance between the two copper electrodes.

.... **(1 mark)**

I–V characteristics

1 Images A–D in the figures below represent the *I–V* characteristics of four components. Which of A–D are ohmic conductors? **(2 marks)**

☐ **A** ☐ **B** ☐ **C** ☐ **D**

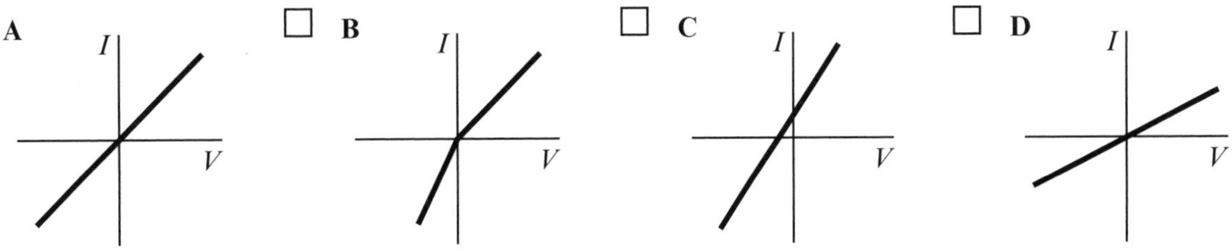

2 The graph shows part of the *I–V* characteristic of a filament lamp.

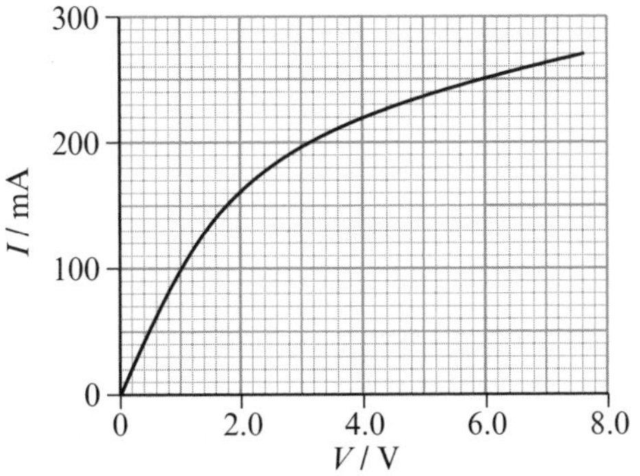

(a) Determine the resistance of the lamp when the potential difference across it is 2.0 V.

....

....

(2 marks)

(b) Determine the resistance of the lamp when the potential difference across it is 6.0 V.

....

.... **(1 mark)**

Guided

(c) Account for any difference between your answers for (a) and (b) above.

The resistance at 2.0 V, (0.16 A) is lower than the resistance at 6.0 V, (0.25 A) because as the current increases, the temperature of the filament increases ....

....

.... **(2 marks)**

3 The graph shows part of the *I–V* characteristic of a diode.

If you can include quantitative descriptions, you should do.

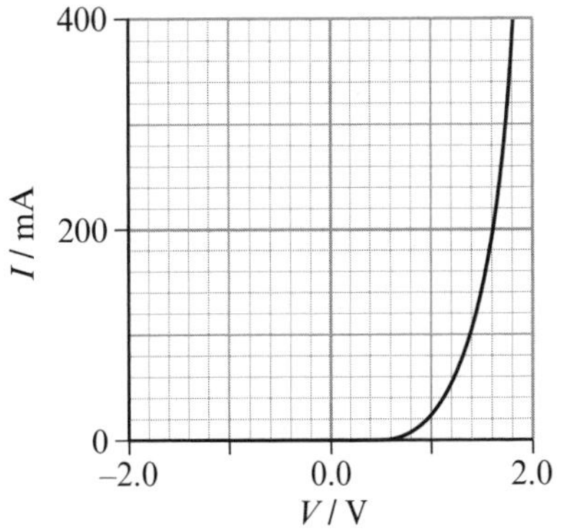

Describe in detail how the behaviour of the diode changes as the potential difference across it varies from –2.0 V to 2.0 V.

....

....

....

.... **(3 marks)**

Had a go ☐ Nearly there ☐ Nailed it! ☐

Resistance and resistivity

1 The heating element in a hairdryer consists of a coil of nickel–chrome alloy wire. The element is connected to 230 V mains and uses a current of 4.35 A. The resistivity of the nickel–chrome wire is $1.06 \times 10^{-6}\,\Omega$m.

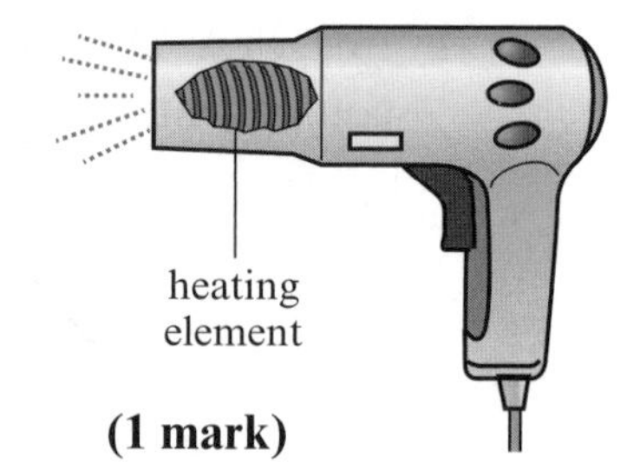

(a) Determine the resistance of the heating element.

.... **(1 mark)**

Guided (b) Determine the length of wire required to construct the heating element.

$R = \rho l/A$ can be rearranged to give $l = \frac{RA}{\rho}$ = ...

.... **(3 marks)**

Guided **2** A strain gauge consists of a network of fine wires made of an alloy such as constantan attached to a small piece of plastic film (see figure). The strain gauge can be attached to a metal component, so when the latter is deformed when stressed, the resistance of the strain gauge will change as the dimensions of the fine wires also change. The change in resistance indicates how much the component has been deformed.

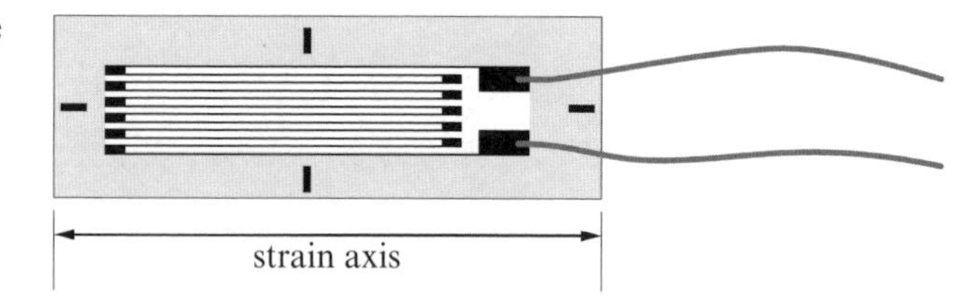

The above strain gauge consists of 12 wires, each 10 mm long by 0.10 mm wide by 5.0 μm thick. The resistivity of constantan is $4.9 \times 10^{-7}\,\Omega$m.

(a) Calculate the resistance of the strain gauge.

5.0 μm = 5.0×10^{-6} m and the twelve wires are in series so using

$R = \frac{\rho l}{A}$ gives ....

.... **(3 marks)**

(b) During a particular test, the strain gauge is subjected to a strain of 0.1% along its axis. If the volume of the wires that make up the strain gauge remains constant, show that the cross-sectional area of the wires will decrease by about 0.1%.

....

.... **(2 marks)**

Guided (c) Explain why the resistance of the strain gauge would be expected to increase by about 0.2% when it is subjected to 0.1% strain.

The above result (b) tells us that the wire in the strain gauge gets 0.1% longer and 0.1% thinner in terms of area of cross-section.

$R = \frac{\rho l}{A}$ so we know that both of these effects increase the resistance.

....

.... **(2 marks)**

Resistivity and temperature

1 In an experiment to investigate how the resistivity of copper varies with temperature, the circuit shown below is used.

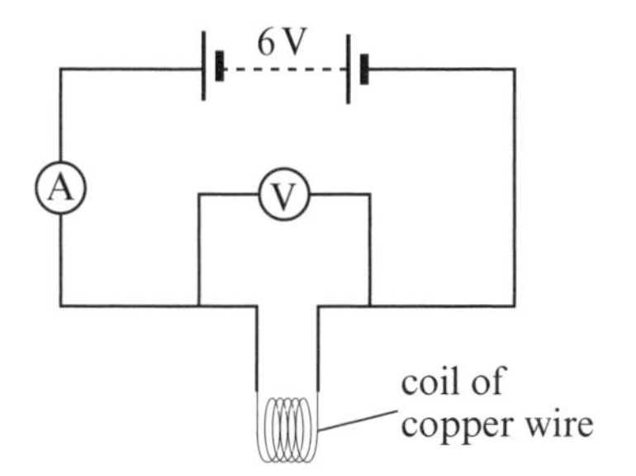

T / °C	*I* / A	*V* / V	*R* / Ω
11	1.20	1.38	1.15
20	1.18	1.38	1.17
32	1.12	1.39	1.24
41	1.09	1.39	1.28
50	1.05	1.39	1.32
59	1.02	1.40	

The coil of copper wire is immersed in a water bath and a series of measurements of current and p.d. are taken over a range of temperatures. The data obtained are listed in the table to the right of the circuit diagram.

(a) Complete the final blank cell in the data table. **(1 mark)**

(b) Using the blank grid below, plot a graph of resistance *R* against temperature *T* and add a line of best fit. **(5 marks)**

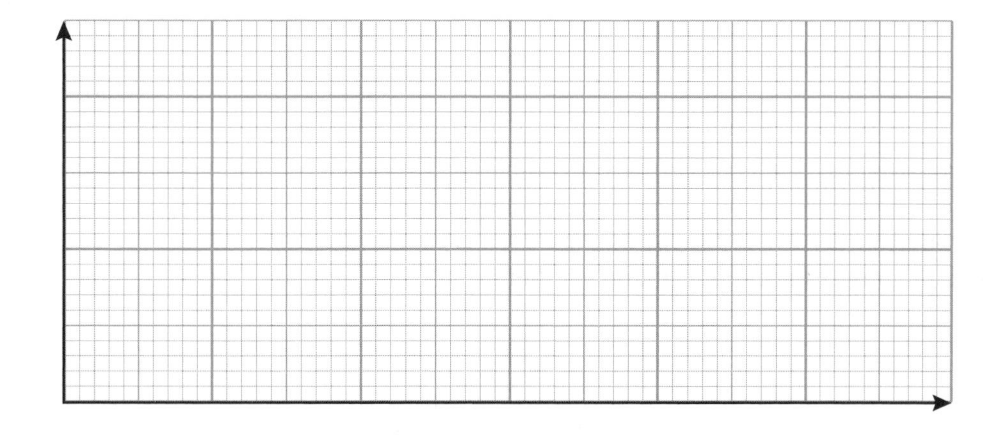

Choose the scales carefully. The resistance only changes by a small amount so it would not be sensible to start at zero ohms.

(c) Determine the gradient of the line to an appropriate precision including its units.

.... **(2 marks)**

(d) What would the resistance of the wire be at 0 °C?

.... **(1 mark)**

(e) Calculate the temperature at which extrapolation of the data and the graph suggest that the resistance of the coil of copper wire would drop to zero.

....

.... **(2 marks)**

2 The temperature of an NTC thermistor is increased from 20 °C to 40 °C. The most likely effect on its resistance is that the resistance would:

☐ **A** double ☐ **C** halve

☐ **B** increase but not double ☐ **D** decrease but not halve. **(1 mark)**

Electrical energy and power

1 Electrons are accelerated in a cathode ray tube between the cathode and the anode by a potential difference of 5 kV.

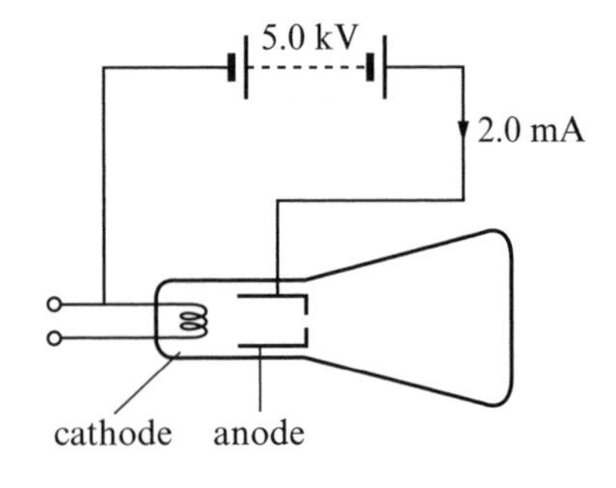

Guided

(a) Given the elementary charge, $e = 1.60 \times 10^{-19}$ C, determine the kinetic energy gained by one electron as it accelerates between the cathode and the anode.

The kinetic energy of the electron is equal to the work done by the accelerating voltage so we can use $E_k = W = VQ =$ ....

.... **(2 marks)**

Guided

(b) The mass of an electron is $m_e = 9.11 \times 10^{-31}$ kg. Determine the velocity of an electron that has been accelerated between the cathode and the anode.

$E_k = \frac{1}{2}mv^2$ so we can find v by rearranging this equation to give

$v = \sqrt{\frac{2E_k}{m}}$ ....

.... **(2 marks)**

(c) Define the quantity **power**.

....

.... **(1 mark)**

(d) The electron beam in the cathode ray tube carries a current of 2 mA. Determine the power required to maintain the beam.

....

.... **(2 marks)**

(e) The cathode consists of a tungsten wire filament through which a current of 0.30 A is passed from an additional power supply. The resistance of the filament is 21 Ω. Determine the electrical power required by the filament.

....

.... **(2 marks)**

2 An electrical heater for use in the home is powered by 230 V electrical mains and has an output power of 2000 W. The heater has a thermostat that switches the heater off when the room temperature reaches 20 °C. This means that, on average, the heater is only on for 45 minutes for every hour that it is in use.

Electricity costs 9.5 pence per kW h. Determine the cost of using the heater for 6 hours.

....

....

.... **(3 marks)**

Kirchhoff's laws and circuit calculations

1 In electrical circuits, the principle of conservation of energy is expressed in:

☐ **A** Kirchhoff's first law ☐ **C** Kirchhoff's second law

☐ **B** Ohm's law ☐ **D** none of the above. **(1 mark)**

2 In the following circuits, each cell has an e.m.f. of 1.5 V and each resistor has a resistance of 5.0 Ω.

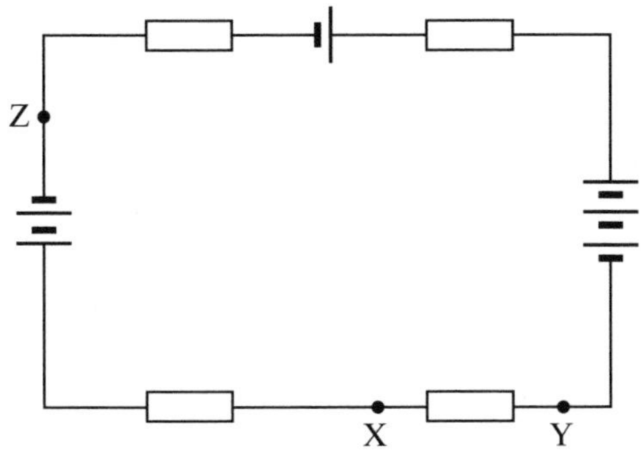

Z

X Y

Make sure you carefully check the orientation of the cell and batteries and then add or subtract their e.m.f.s accordingly.

Figure 1

(a) In Figure 1, the potential difference between X and Y is:

☐ **A** 2.3 V ☐ **B** zero ☐ **C** 6.0 V ☐ **D** 1.5 V **(1 mark)**

(b) The current around the circuit at Z is:

☐ **A** 0.45 A ☐ **B** 0.30 A ☐ **C** 0.15 A ☐ **D** zero **(1 mark)**

In the circuit in Figure 2, the current in resistor C is 0.5 A:

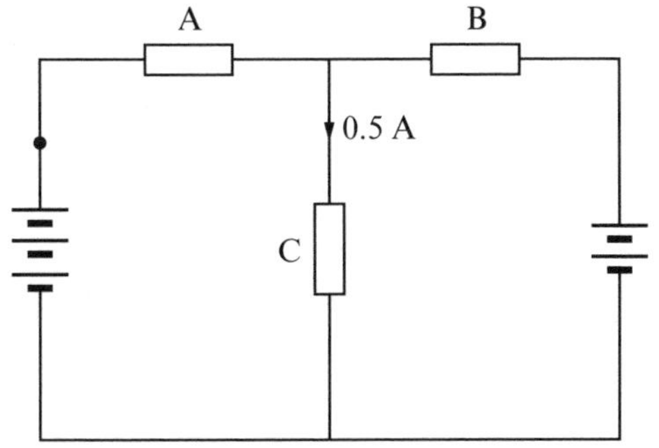

A B

0.5 A

C

Figure 2

(c) The current in resistor A is:

☐ **A** 0.9 A to the right ☐ **C** 0.15 A to the right

☐ **B** 0.5 A to the right ☐ **D** 0.4 A to the right. **(1 mark)**

(d) The current in resistor B is:

☐ **A** 0.1 A to the left ☐ **C** 0.9 A to the right

☐ **B** 0.15 A to the left ☐ **D** 0.4 A to the right. **(1 mark)**

Had a go ☐ Nearly there ☐ Nailed it! ☐

Resistors in series and parallel

1 The resistors in the arrangements below are all identical.

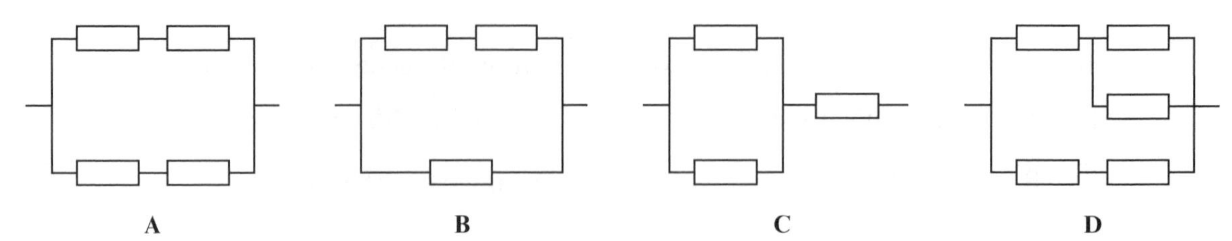

(a) Which of A–D has the highest resistance? .... **(1 mark)**

(b) Which of A–D has the lowest resistance? .... **(1 mark)**

Guided

2 You have **five** 15 Ω resistors. Draw diagrams to show how would you produce the following resistances using **some** or all of your resistors.

(a) 20 Ω

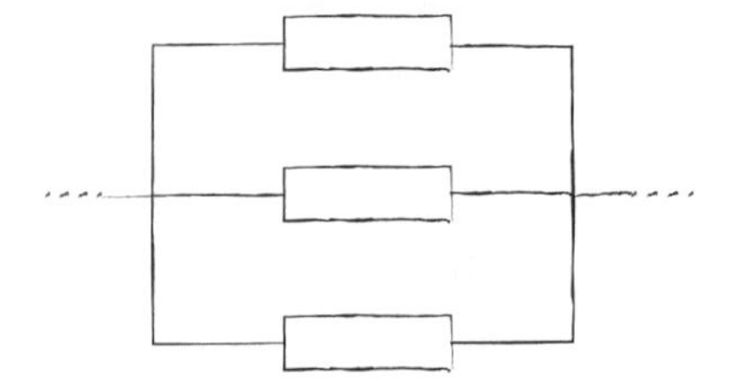

Three 15 Ω resistors in parallel will provide a useful 5 Ω to start you off for (a) as shown here. You only have five 15 Ω resistors, so you will need to think about (b) a bit more. 10 Ω is less than 15 Ω so a parallel arrangement is involved again.

(1 mark)

(b) 10 Ω

(1 mark)

3 Determine the resistances of the following combinations of resistors.

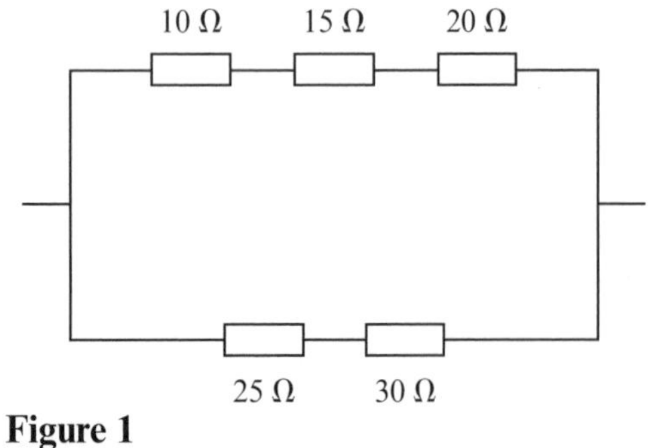

Figure 1

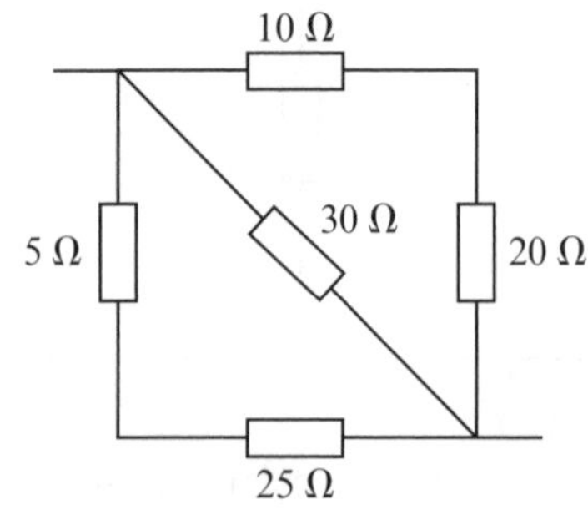

Figure 2

(a) Figure 1

....

.... **(1 mark)**

(b) Figure 2

....

.... **(1 mark)**

DC circuit analysis

1 Figure 1 shows the $\frac{I}{V}$ characteristic of a red light-emitting diode (LED). It is to be used in the circuit in Figure 2.

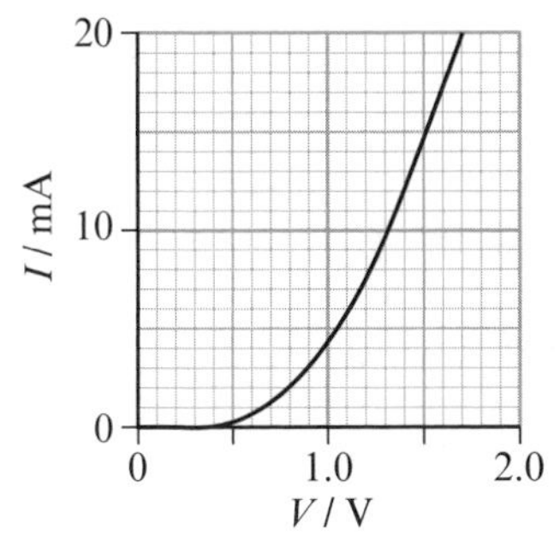

Figure 1

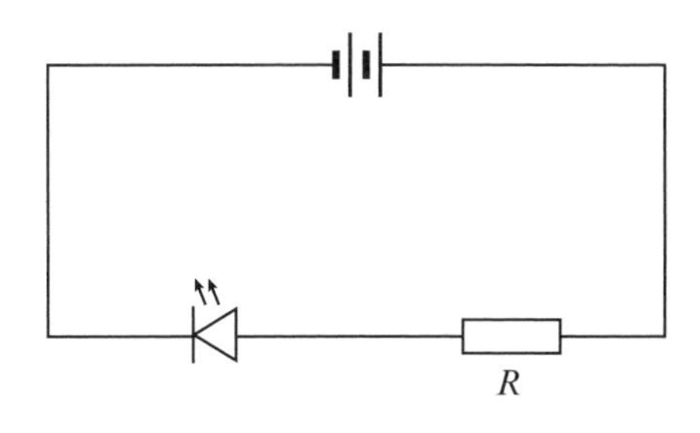

Figure 2

(a) State the p.d. across the LED when the current through it is 20 mA.

.... **(1 mark)**

(b) The e.m.f. of the battery is 3.0 V. Its internal resistance is negligible. Determine the value of R that will result in a current of 20 mA flowing through the LED when it is used in the circuit of Figure 2.

....

.... **(2 marks)**

(c) After the circuit has been running for a period of time, the LED is dimmer because the e.m.f. of the battery has decreased. The p.d. across the diode is now 1.1 V. Determine the value to which the e.m.f. of the battery has fallen.

....

....

.... **(2 marks)**

(d) Six of the above red LEDs are connected together in an array as shown below:

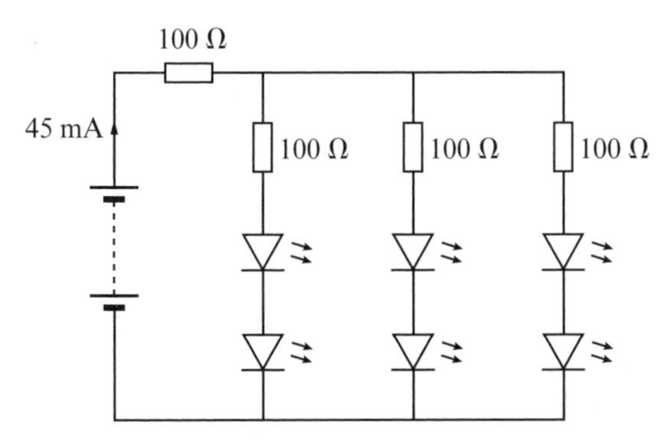

Determine the e.m.f. of the battery in the above circuit.

....

....

....

....

.... **(3 marks)**

Had a go ☐ Nearly there ☐ Nailed it! ☐

E.m.f. and internal resistance

Guided

1 A battery of e.m.f. ε and internal resistance r is connected across an external resistor R as shown:

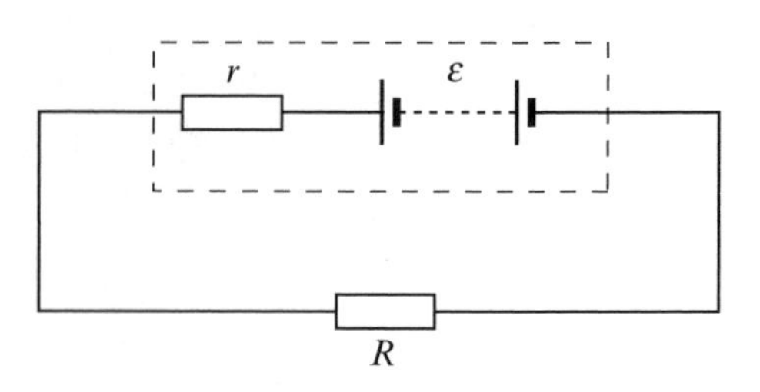

If the potential difference across R is V, show that:

$$V = \varepsilon \frac{R}{R + r}$$

The current in the circuit is the e.m.f. of the battery divided by the total circuit resistance including the internal resistance and it is also equal to the terminal voltage divided by the external resistance.

....

....

.... **(2 marks)**

2 All the cells in this question have an e.m.f. of 1.5 V and an **internal resistance** of 0.2 Ω.

The internal resistance can be treated just like a normal resistor in series with each cell.

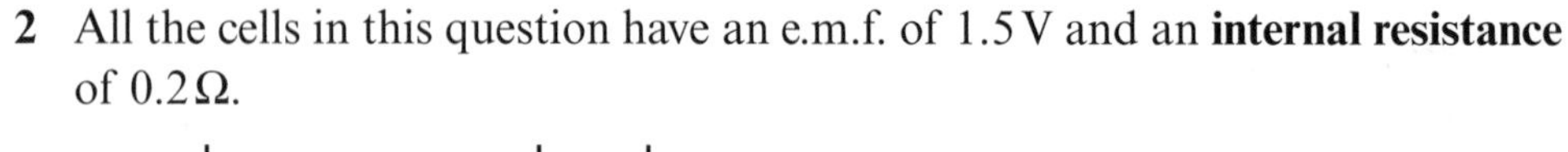

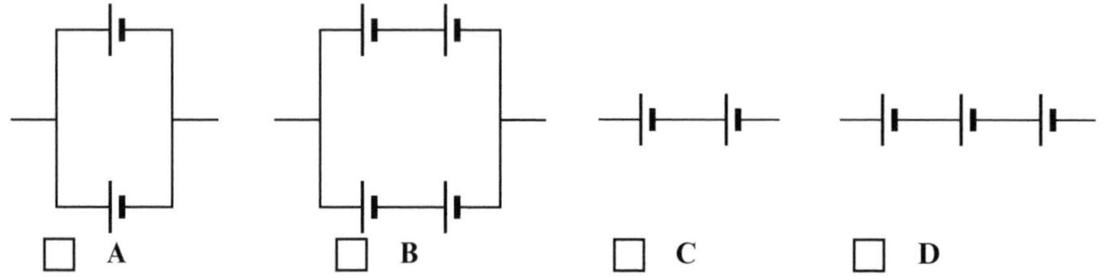

☐ **A** ☐ **B** ☐ **C** ☐ **D**

Which of the batteries A–D in the figures above will provide the most current when connected across a 0.5 Ω resistor? .... **(1 mark)**

3 A 6.0 V battery with internal resistance 0.50 Ω is connected across a filament lamp which is **rated** at 6 V, 8 W (see figure). The current around the circuit is measured and found to be 1.2 A.

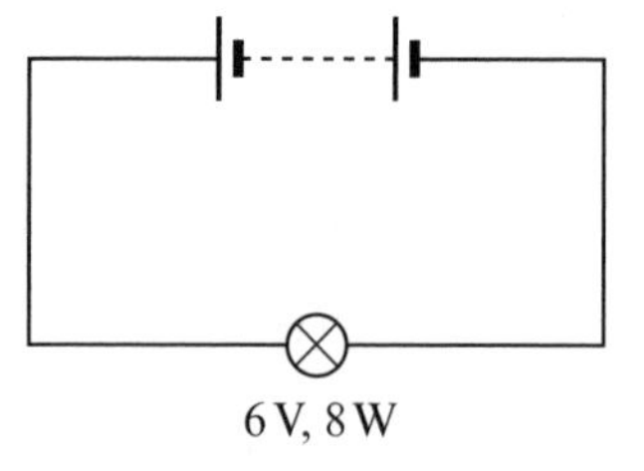

(a) Determine the p.d. across the internal resistance of the battery.

.... **(1 mark)**

(b) Determine the p.d. across the lamp.

.... **(1 mark)**

(c) Determine the power dissipated by the lamp.

.... **(1 mark)**

(d) Explain why the lamp cannot light with full brightness when connected in this circuit.

....

....

.... **(2 marks)**

Experimental determination of internal resistance

1 The circuit in the figure is used to determine the internal resistance of a single cell.

The variable resistor allows the current provided by the cell to be varied. The current and the p.d. across the cell are measured using a digital ammeter (0–9.99 A) and digital voltmeter (0–19.99 V), respectively.

(a) State the uncertainty in the measurements of the current and the p.d.

....

....

The meters are digital, so the uncertainty is a full division, not half of one.

(1 mark)

(b) Using the data below, plot a graph of the p.d. across the cell, V, against the current, I, through the cell on the axes below. Include a line of best fit and error bars.

I / A	V / V
0.00	1.45
0.10	1.44
0.20	1.44
0.30	1.42
0.39	1.41
0.51	1.41

(5 marks)

(c) Explain why the gradient of the graph is equal to $-r$, where r is the internal resistance of the cell.

....

....

.... **(2 marks)**

(d) Determine the gradient of the graph and the internal resistance of the cell.

.... **(2 marks)**

(e) What is the e.m.f. of the cell?

.... **(1 mark)**

(f) How might the apparatus chosen or the measurements taken have been improved in order to reduce the uncertainty in the determination of r?

....

....

....

.... **(2 marks)**

Had a go ☐ Nearly there ☐ Nailed it! ☐

Potential dividers

1 Consider this circuit.

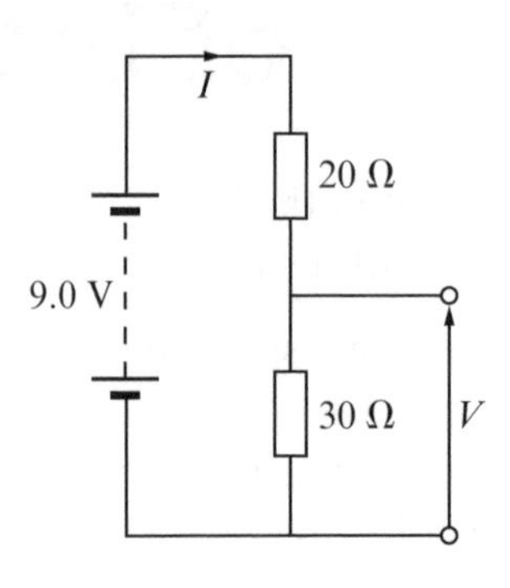

Guided

(a) Calculate the potential difference V across the 30 Ω resistor.

The potential divider formula $V_{out} = V_{in}\dfrac{R_2}{R_1 + R_2}$ gives ....

....

.... **(2 marks)**

(b) Now consider this circuit:

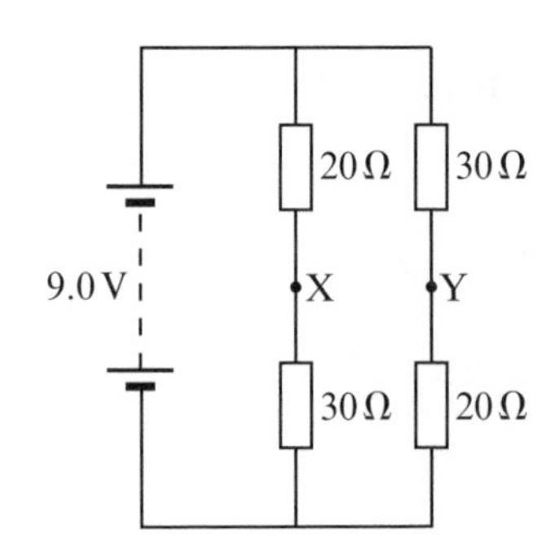

Determine the p.d. between points X and Y.

....

.... **(2 marks)**

2

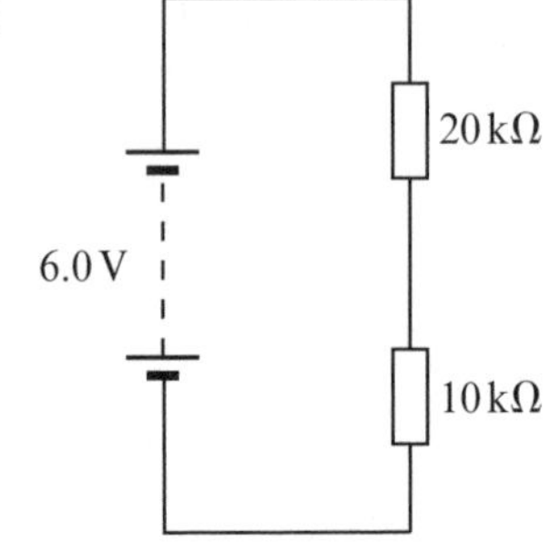

Figure 1

Figure 2

(a) Determine the p.d. across the 10 kΩ resistor in the circuit in Figure 1.

....

.... **(1 mark)**

(b) A voltmeter that also has a resistance of 10 kΩ is connected across the 10 kΩ resistor as in the circuit in Figure 2. What is the reading on the voltmeter?

....

....

.... **(2 marks)**

Investigating potential divider circuits

1 An NTC thermistor has the resistance–temperature characteristic shown in this graph.

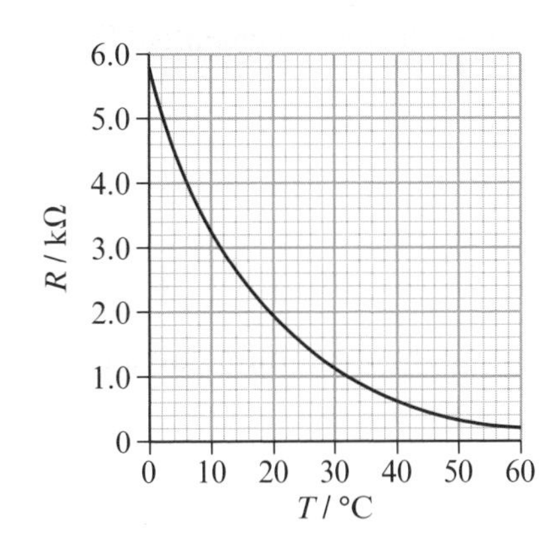

The sensitivity is how much the resistance changes per degree in $\Omega\,°C^{-1}$.

(a) How does the sensitivity of the thermistor vary over the range 0–60 °C?

.... **(1 mark)**

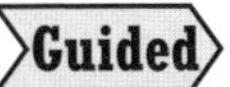

(b) Determine the sensitivity of the thermistor at 30 °C.

The sensitivity of the thermistor at 30 °C is equal to the slope of the curve at 30 °C so a tangent to the curve ....

....

.... **(3 marks)**

The thermistor is used in the following circuit in order to produce a temperature sensor with an output p.d. that increases with increasing temperature.

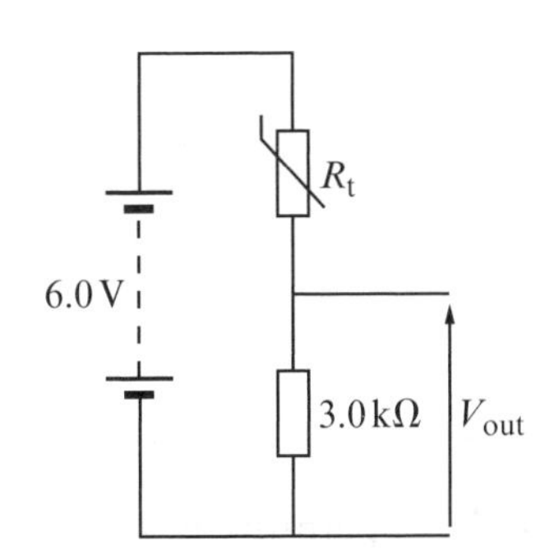

The resistance is given in kΩ, but as this is a potential divider, you need not convert to Ω provided you remember that all resistances are in kΩ.

(c) What is the output p.d., V_{out}, when the temperature of the thermistor is 12 °C?

....

.... **(2 marks)**

(d) What is the output p.d. when the temperature of the thermistor is 40 °C?

....

....

.... **(2 marks)**

(e) At what temperature will the output p.d. be 3.6 V?

....

....

.... **(2 marks)**

Had a go ☐ Nearly there ☐ Nailed it! ☐

Exam skills

1 A battery has electromotive force 12.0 V and an internal resistance of 1.0 Ω.

(a) Explain the term **electromotive force**.

....

....

.... **(2 marks)**

(b) Determine the maximum current that the battery can supply.

....

.... **(1 mark)**

The battery is connected across a lamp as shown.
The p.d. across the lamp is 10.0 V.

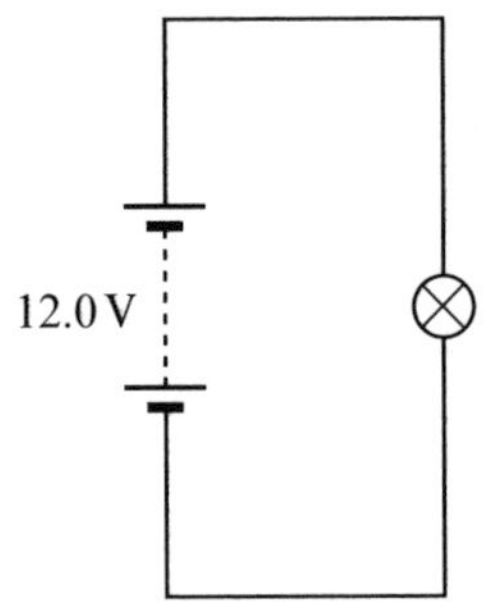

12.0 V

(c) State the p.d. across the battery terminals.

.... **(1 mark)**

(d) Account for the difference between the e.m.f. of the battery and the p.d. across its terminals.

....

....

....

.... **(2 marks)**

(e) Given that lamp A is operating normally, determine current supplied by the battery.

....

....

.... **(2 marks)**

A second identical lamp is added to the circuit as shown:

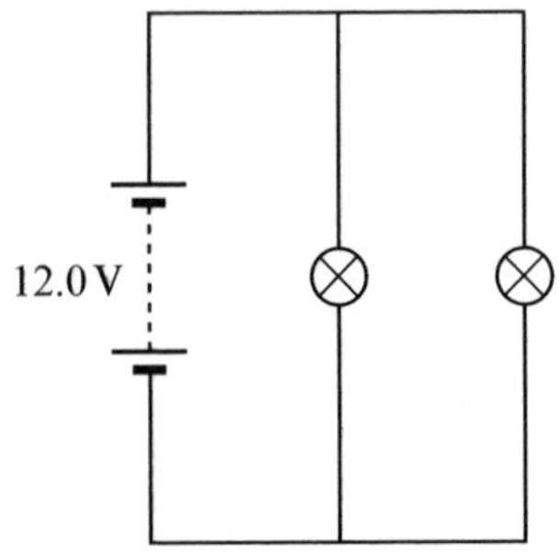

12.0 V

(f) Explain why both lamps light but with less than normal brightness.

....

....

....

.... **(3 marks)**

Properties of progressive waves

1 The figure below represents a 'snapshot' of a progressive wave.

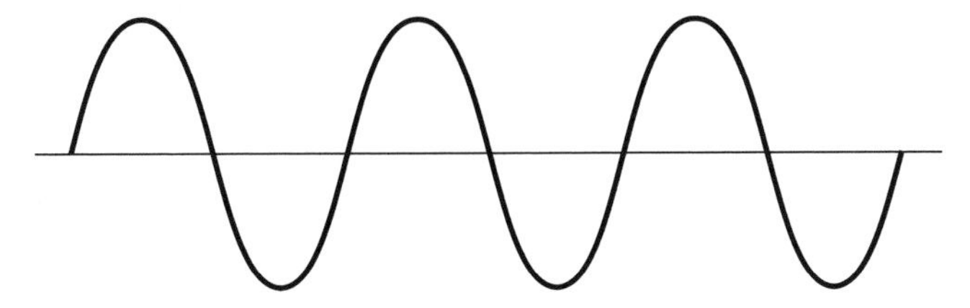

(a) Add labels to the diagram to indicate a **crest** and a **trough**. **(2 marks)**

(b) Identify and clearly label the **amplitude** of the wave and the **wavelength** of the wave. **(2 marks)**

Guided

2 Describe the difference between a **transverse** wave and a **longitudinal** wave. You should use diagrams to aid your explanation.

In the case of transverse waves, the particles carrying the wave oscillate at right angles to the direction of propagation of the wave or the direction of energy transfer. For longitudinal waves ....

....

....

....

.... **(4 marks)**

3 The figure represents two progressive waves, A and B.

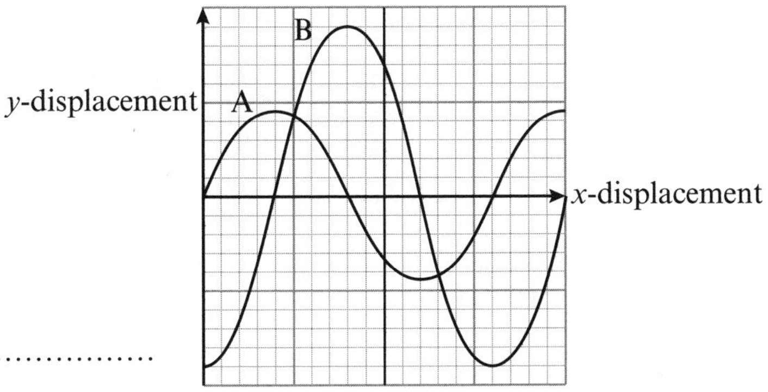

y-displacement

A

B

x-displacement

(a) Determine the phase difference between wave A and wave B.

....

....

.... **(2 marks)**

(b) Determine the ratio of the intensity of wave B to the intensity of wave A.

....

....

Intensity is proportional to amplitude squared.

(2 marks)

Had a go ☐ Nearly there ☐ Nailed it! ☐

The wave equation

Guided

1 A radio station transmits radio waves with a frequency of 96.1 MHz. A simple receiving aerial on a radio consists of a telescopic metal tube with a length equal to one quarter of the wavelength of the radio waves it is designed to receive.

Given that radio waves travel at $3.00 \times 10^8\,\text{m}\,\text{s}^{-1}$, determine the length of aerial required to receive a 96.1 MHz signal.

Rearranging $v = f\lambda$ gives $\lambda = \frac{v}{f}$ ....

....

.... **(2 marks)**

2 Ultrasound waves with a wavelength of 0.44 mm are used for medical imaging purposes. The speed of ultrasound waves in soft tissue is typically around $1540\,\text{m}\,\text{s}^{-1}$.

(a) Determine the frequency of such ultrasound waves.

....

.... **(2 marks)**

(b) What is the advantage of using ultrasound waves with a very short wavelength?

....

....

Think about the effect of wavelength on diffraction and the ability to observe fine detail.

....

.... **(2 marks)**

(c) Give one advantage of using ultrasound for medical imaging rather than X-rays.

....

.... **(1 mark)**

3 A green laser pointer produces light with a wavelength of 532 nm in air.

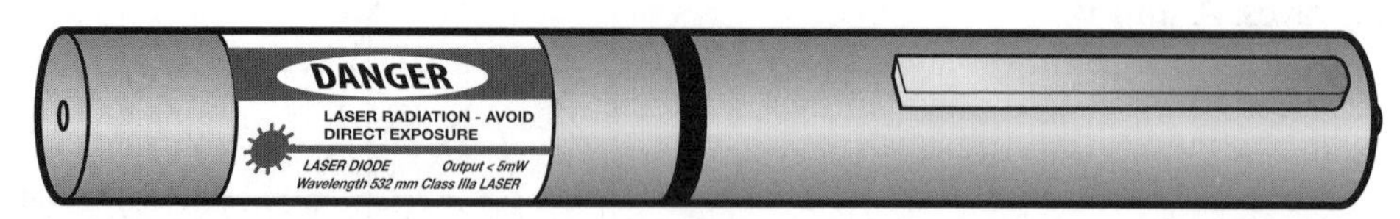

(a) The speed of light in air is $3.00 \times 10^8\,\text{m}\,\text{s}^{-1}$. Determine the frequency of the green light.

....

.... **(2 marks)**

(b) When light enters water from air, it is slowed down by a factor of 1.33, the refractive index of water. Complete the table for light from the laser pointer.

	In air	In water
Speed / $\text{m}\,\text{s}^{-1}$	3.00×10^8	
Wavelength / nm	532	
Frequency / Hz		

(3 marks)

Graphical representation of waves

1 The graphs below show how the particle displacement y varies with position x along a wave and at time t, respectively:

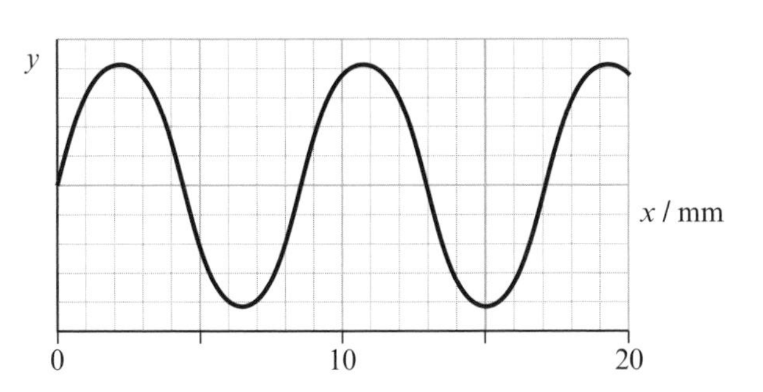

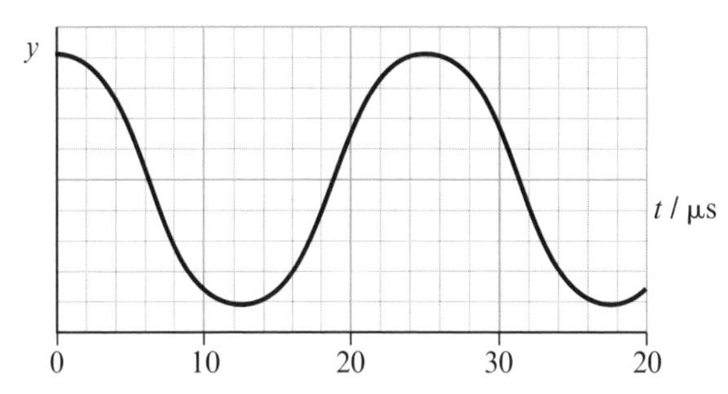

(a) Determine the wavelength of the wave.

....

.... **(1 mark)**

(b) Determine the frequency of the wave.

....

....

.... **(2 marks)**

(c) Determine the speed of the wave.

....

.... **(1 mark)**

2 (a) On the blank grid below, sketch a wave with an amplitude of 1.5 mm and a wavelength of 20 mm. **(2 marks)**

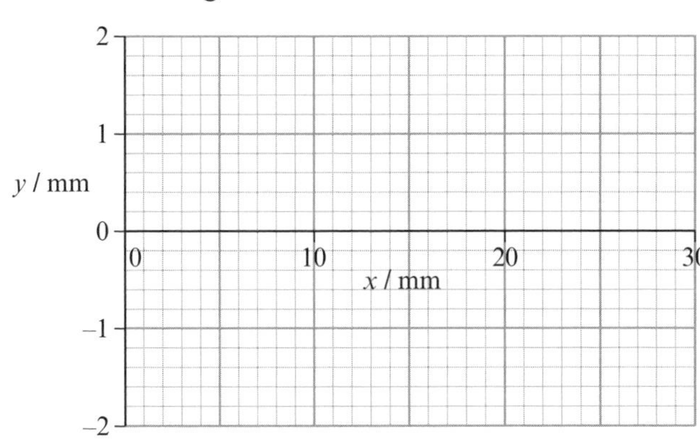

Remember that amplitude is measured from zero displacement and not from peak to trough.

(b) On the same grid, sketch a second wave of the same amplitude and wavelength but differing in phase by $\frac{\pi}{3}$ radians. **(1 mark)**

One complete wave is equivalent to a phase of 2π radians (think of the diameter of a circle), so $\pi/3$ is equivalent to one sixth of a wave.

Had a go ☐ Nearly there ☐ Nailed it! ☐

Using an oscilloscope to display sound waves

1 An electrical signal is displayed on the screen of an oscilloscope as shown below:

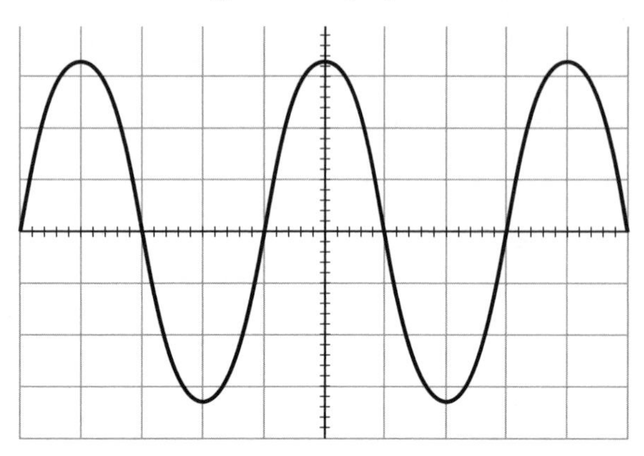

The vertical (y) sensitivity is set to 1 V/ division and the time base is set to 1 ms / division.

(a) Name a device that can convert a sound signal into an electrical signal.

.... **(1 mark)**

(b) Determine the amplitude of the signal in the figure above.

....

.... **(2 marks)**

Guided

(c) Determine the frequency of the signal in the figure above.

One complete wave occupies 4 divisions or 4 ms, which is equal to

0.004 s ....

....

.... **(2 marks)**

(d) On the blank grid below, sketch the appearance of the same waveform if the time base is changed to 500 µs/div and the vertical (y) sensitivity is set to 2 V / div.

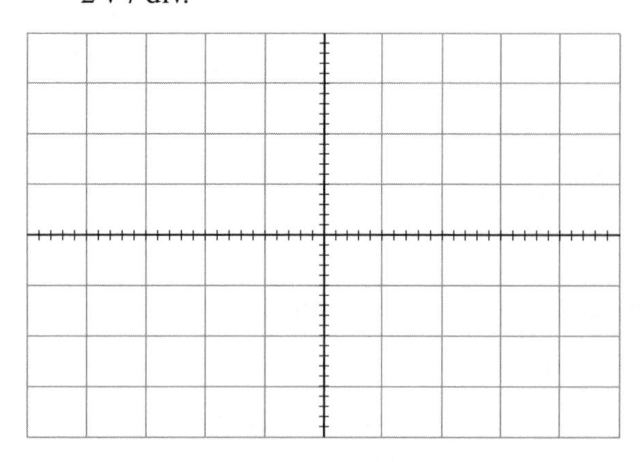

(3 marks)

Reflection, refraction and diffraction

1 The diagram below shows a ripple tank seen from above.

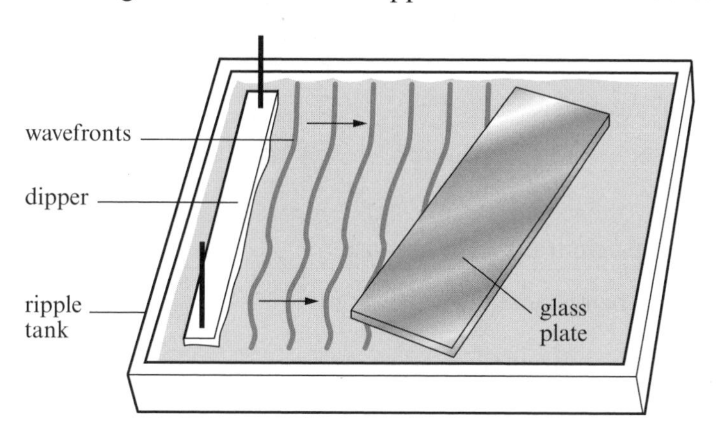

The dipper produces plane waves that pass over a submerged glass plate. The waves travel more slowly when the water is shallower.

(a) What effect does reducing the water depth have on the frequency of the ripples?

.... **(1 mark)**

(b) What effect does reducing the water depth have on the wavelength of the ripples?

.... **(1 mark)**

(c) Wavefronts are shown travelling from the dipper towards the glass plate. Add to the diagram above to show how the wavefronts propagate over the glass plate. Indicate the direction that the ripples travel over the glass plate. **(3 marks)**

(d) Complete the diagram above to show how the ripples propagate once they have passed over the glass plate. **(2 marks)**

The same ripple tank is now arranged so that ripples are directed through a gap between two barriers (see figure below):

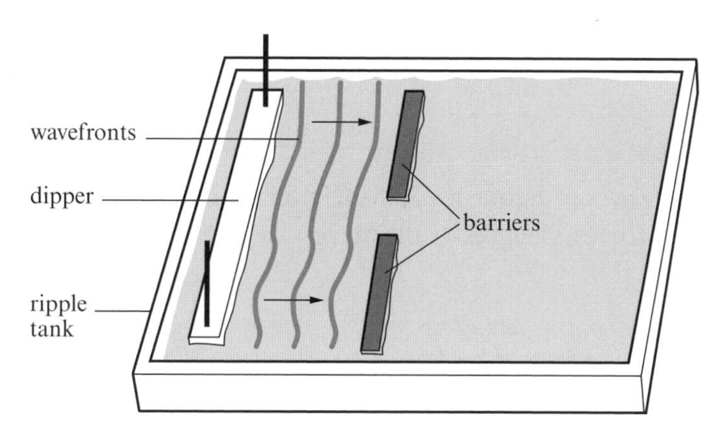

(e) Add to the diagram to show how the ripples behave as they emerge from the gap to the right of the barriers. **(2 marks)**

(f) Name the physical process that the ripples are demonstrating as they emerge from the gap.

.... **(1 mark)**

Had a go ☐ Nearly there ☐ Nailed it! ☐

The electromagnetic spectrum

1 Electromagnetic waves with a wavelength of 100 nm would be:

☐ **A** X-rays ☐ **C** visible light

☐ **B** ultraviolet ☐ **D** infrared. **(1 mark)**

You need to know at least the orders of magnitude of the wavelengths of the different sections of the EM spectrum.

2 Microwaves have a wavelength of the order of:

☐ **A** 10 nm ☐ **B** 10 cm ☐ **C** 10 m ☐ **D** 10 km **(1 mark)**

3 We know that electromagnetic waves are transverse because:

☐ **A** electromagnetic waves can be refracted

☐ **B** electromagnetic waves can be diffracted

☐ **C** electromagnetic waves can travel in a vacuum

☐ **D** electromagnetic waves can be polarised. **(1 mark)**

4 Which of the following is a form of ionising radiation?

☐ **A** ultrasound ☐ **C** ultraviolet

☐ **B** microwaves ☐ **D** infrared. **(1 mark)**

Guided

5 A lamp emits light uniformly in every direction. The power of the lamp is 100 W, and it can be treated as a point source of light. If the efficiency of the lamp is 7.5%, what is the intensity of the light produced by the lamp measured at a distance of 2.0 m from the lamp?

The area of a sphere of radius r is given by $A = 4\pi r^2$ and intensity is power/area ....

....

....

.... **(3 marks)**

6 A microwave oven produces microwaves of frequency 2.45 GHz. The internal cavity of the oven has a width equal to three wavelengths of the microwave radiation.

(a) What is the width is the cavity?

....

.... **(2 marks)**

(b) The microwaves set up a standing wave pattern. Why does this necessitate the use of a rotating plate inside the oven?

....

....

.... **(2 marks)**

Polarisation

1 Which of the following waves cannot be polarised?

☐ **A** ultraviolet ☐ **B** X rays ☐ **C** ultrasound ☐ **D** radio **(1 mark)**

Guided

2 The figures below shows possible arrangements of two polarising filters.

(a) (b) (c)

(a) Explain why a single polarising filter will transmit 50% of incident unpolarised light.

The direction of polarisation of light is actually determined by a vector quantity. Unpolarised light oscillates in every direction but can be resolved into two perpendicular components. ....

....

.... **(2 marks)**

(b) A second polarising filter is placed on top of the first filter and at 45° to it. Explain why some light will pass through the second filter.

....

....

Think again about vectors. There is a component of the polarised light parallel to the transmission direction of the filter.

.... **(2 marks)**

(c) The second filter is rotated until it is at 90° to the first filter. Explain why no light will emerge from the second filter.

....

.... **(2 marks)**

3 Describe an experiment to show that a laboratory microwave transmitter emits polarised microwave radiation. A diagram of the apparatus can be used to aid your answer.

....

.... **(4 marks)**

Had a go ☐ Nearly there ☐ Nailed it! ☐

Refraction and total internal reflection of light

1 The speed of light in air is $3.00 \times 10^8\,m\,s^{-1}$. Determine the speed of light in the core of an optical fibre with a refractive index equal to 1.45.

....

.... **(1 mark)**

2 A 45° prism is made from acrylic plastic that has a refractive index equal to 1.49. A ray is incident, as shown.

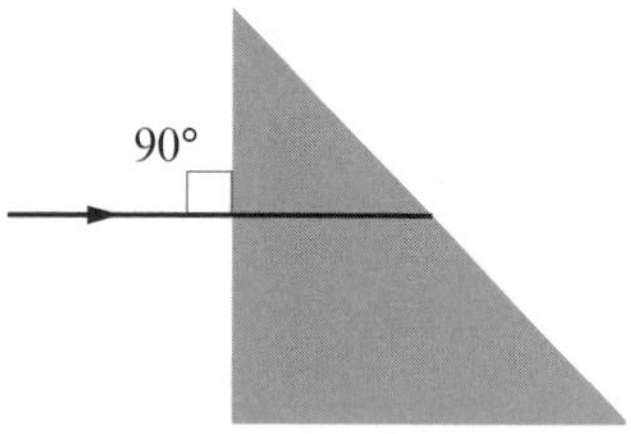
90°

Show that total internal reflection can occur at the sloping face of the prism.

....

....

.... **(2 marks)**

3 A ray of light enters a glass tank of water at an angle of 40° (see figure on the right).

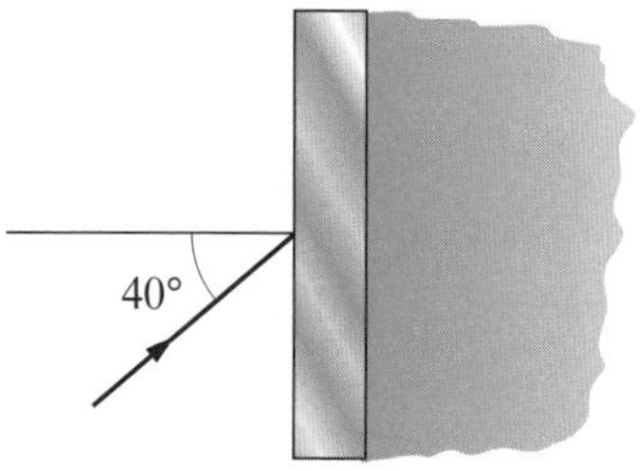
40°
air
$n = 1.00$
glass
$n = 1.52$
water
$n = 1.33$

Guided (a) Determine the angle of refraction at the air–glass boundary.

As $n \sin \theta$ = constant, at the air–glass boundary we can say that $n_{air} \sin 40° = n_{glass} \sin \theta_g$ ....

....

.... **(2 marks)**

(b) Determine the angle at which the ray enters the water.

....

....

.... **(2 marks)**

(c) Determine the critical angle at the glass–water boundary.

....

.... **(2 marks)**

(d) Show that total internal reflection cannot occur at the glass–water boundary for a ray at any angle of incidence at the air–glass boundary.

....

....

.... **(3 marks)**

The principle of superposition

1 Two waves in antiphase will produce zero resultant when superposed only if they have the same:

☐ **A** wavelength

☐ **B** frequency

☐ **C** amplitude

☐ **D** velocity. **(1 mark)**

2 When two waves of the same amplitude are in phase, the result of superposition is a wave of:

☐ **A** the same intensity

☐ **B** twice the intensity

☐ **C** four times the intensity

☐ **D** eight times the intensity. **(1 mark)**

3 When two waves of frequency f superpose, the resultant has a frequency of:

☐ **A** f ☐ **C** $2f$

☐ **B** $\sqrt{2}f$ ☐ **D** $4f$ **(1 mark)**

4 Two waves of amplitude A and $2A$ superpose. The resultant **cannot** have amplitude:

☐ **A** A ☐ **C** $3A$

☐ **B** $2A$ ☐ **D** zero **(1 mark)**

5 The figure below is a displacement–time graph of two waves, X and Y.

Add to the diagram the resultant displacement–time graph that would result from the superposition of waves X and Y.

The resultant is found by adding the displacements of the two waves. It easy to work that out when one of the waves has zero displacement.

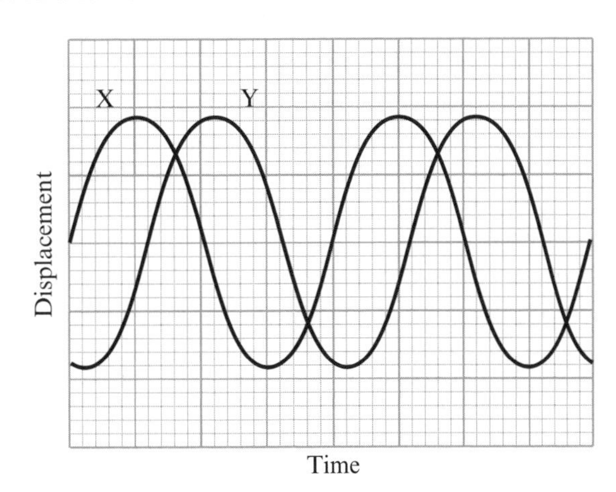

(3 marks)

Interference

Guided

1 A microwave transmitter emits microwaves of wavelength 2.8 cm. The microwaves arrive at a receiver by travelling directly along the line AB and also along the path ACB after reflecting off a metal plate placed at C.

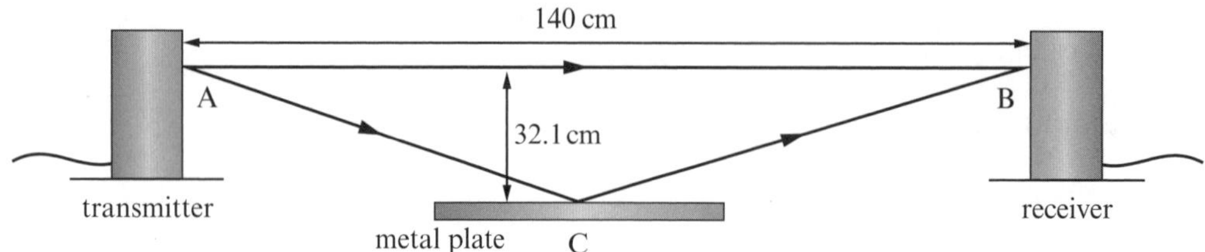

(a) Determine the path length ACB in cm:

Using Pythagoras' theorem: $ACB = 2 \times \sqrt{70.0^2 + 32.1^2}$ ....

....

.... **(2 marks)**

(b) Determine the path difference between ACB and AB:

....

.... **(1 mark)**

Guided

(c) Determine the resultant phase difference between microwaves travelling along path ACB and those travelling directly from A to B:

A path difference of one wavelength results in a phase difference of

2π ....

....

.... **(2 marks)**

(d) State the type of interference at B associated with the phase difference in (c) above.

.... **(1 mark)**

2 Two small lamps are placed 1.5 cm apart in front of a screen, which is placed 1 m away from them as shown in the figure (not to scale). The lamps are connected to an appropriate power supply and glow with normal brightness.

screen

Give, with explanations, two reasons why interference fringes cannot be seen on the screen.

....

....

....

....

.... **(4 marks)**

Two-source interference and the nature of light

1 The Dutch scientist Christiaan Huygens and the English scientist Isaac Newton had quite differing opinions about the nature of light. State the essential principles behind the two scientists' ideas and how they compare with our current understanding.

....

....

....

....

....

....

....

....

.... **(5 marks)**

Guided

2 Describe Young's double-slit experiment and explain how it confirmed aspects of the wave nature of light. A diagram can be used to help with your answer.

lamp

single slit

....

....

....

....

....

.... **(4 marks)**

When a question suggests using a diagram, do so.

3 This question is about **coherence** in relation to sources of light.

(a) Explain the meaning of the term **coherent**.

....

....

.... **(2 marks)**

(b) Explain why only monochromatic sources of light can be coherent.

....

....

.... **(2 marks)**

Had a go ☐ Nearly there ☐ Nailed it! ☐

Experimental determination of the wavelength of light

1 In an experiment to determine the wavelength of the light emitted by a laser pointer, light from the pointer is shone through a double slit in order to produce an interference pattern on a screen some distance away. The diagram of the setup is shown below (not to scale):

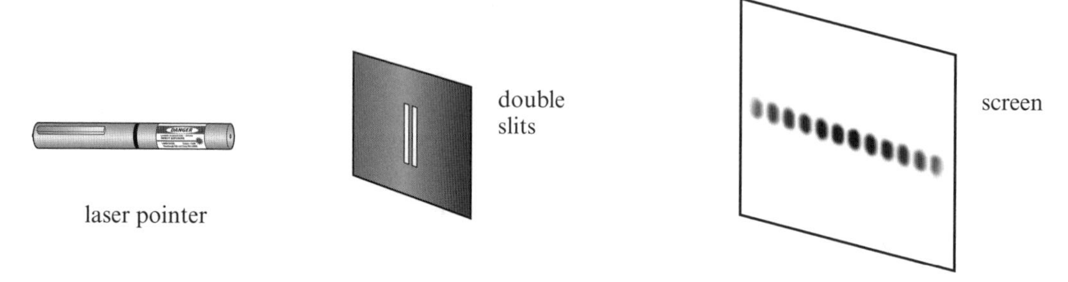

The separation of the slits is 0.20 mm and the screen is placed 3.00 m from the slits. The separation of the bright fringes produced can be measured with a ruler with a mm scale, as shown below:

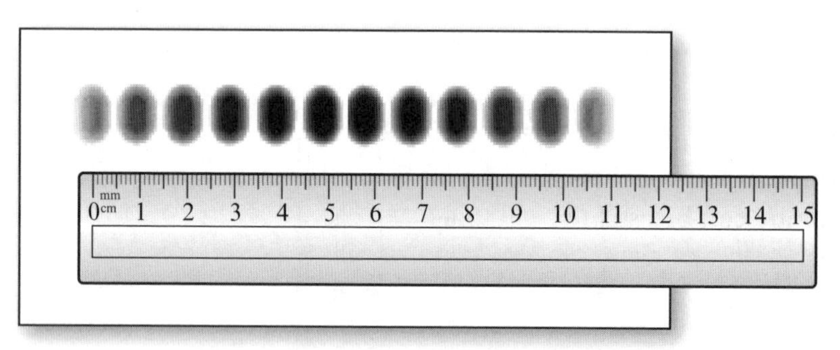

Make sure you count fringe spacings, not fringes.

(a) Use the mm scale to determine the fringe separation.

....

.... **(2 marks)**

(b) Estimate the uncertainty in your measurement of the fringe separation.

....

.... **(2 marks)**

Guided (c) Determine the wavelength of the light emitted by the laser pointer.

Using $\lambda = \frac{ax}{D}$ ....

....

.... **(3 marks)**

Guided (d) The wavelength of the laser pointer is checked by shining its beam through a 600 line per mm diffraction grating. The first-order maximum is found at an angle of 23°. Show that this confirms your answer to part (c) above.

Using $n\lambda = d \sin \theta$ ....

....

.... **(2 marks)**

Stationary waves

1 State the conditions required for the formation of a stationary wave.

....

....

....

....

.... **(4 marks)**

2 The apparatus shown below is to be used to investigate microwave standing waves. The microwave probe can be moved in the space between the microwave transmitter and the metal sheet that acts as a reflector.

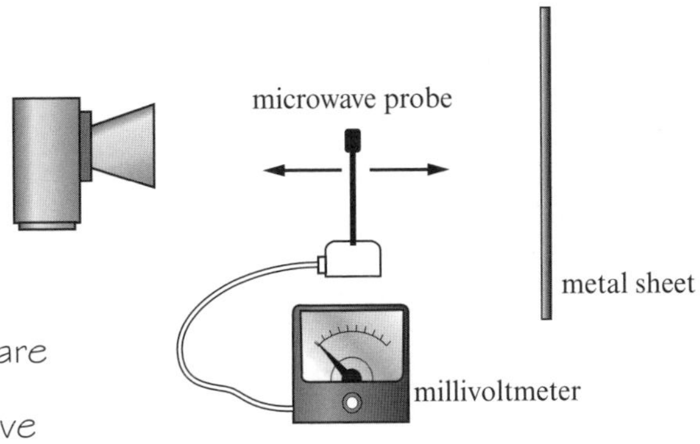

Guided

(a) Explain how microwave standing waves are produced in the space between the microwave transmitter and the metal sheet.

microwave probe

metal sheet

millivoltmeter

Microwaves incident on the metal sheet are reflected, the probe detects a direct wave and a reflected wave giving two waves of similar amplitude but travelling in opposite directions ....

....

....

.... **(4 marks)**

(b) The reading on the millivoltmeter V varies periodically as the probe is moved from the metal sheet towards the transmitter by a distance x as illustrated by the graph.

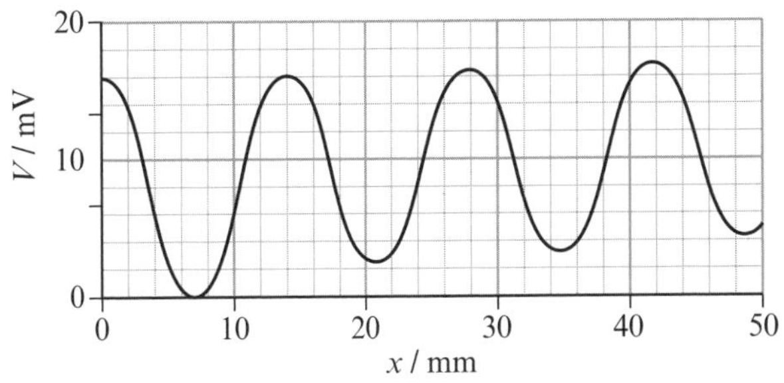

Account for the variation in the signal picked up by the probe.

V / mV: 0, 10, 20

x / mm: 0, 10, 20, 30, 40, 50

....

....

....

.... **(3 marks)**

(c) Determine the wavelength of the microwaves used in the experiment.

....

.... **(2 marks)**

Stationary waves on a string

1 The diagram shows a standing wave of frequency f on a string of length l.

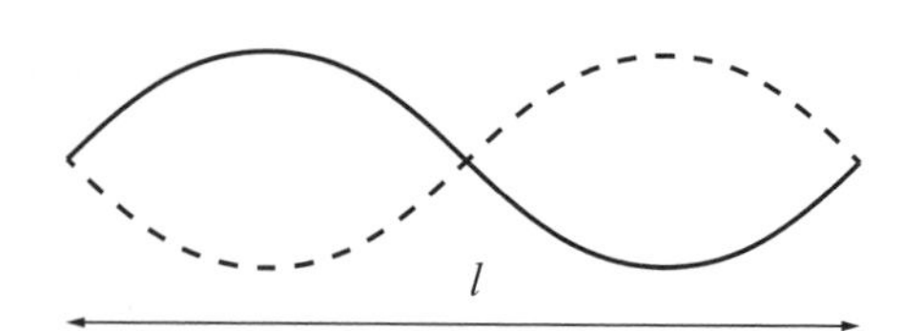

A further standing wave pattern is produced when the frequency is increased to $\frac{3f}{2}$.

(a) Which of the following, A–D, correctly describes the number of nodes and antinodes at frequency $\frac{3f}{2}$?

	Nodes	Antinodes
A	3	3
B	4	4
C	4	3
D	3	4

(1 mark)

(b) Which of the following frequencies A–D will **not** result in standing waves on the above string?

☐ **A** $\frac{f}{2}$ ☐ **B** $2f$ ☐ **C** $\frac{5f}{4}$ ☐ **D** $\frac{7f}{2}$ **(1 mark)**

Guided **2** The figure below shows a standing wave of frequency 16.0 Hz on string of length 1.50 m.

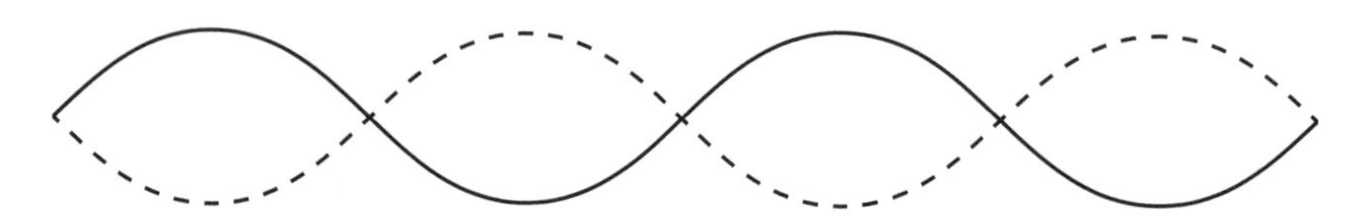

(a) Determine the speed at which waves travel down the string.

The string carries two complete waves so the wavelength is half the length of the string ....

....

....

$v = f\lambda$ applies to waves on strings just like any other wave.

(2 marks)

(b) The tension in the string is increased slowly until another standing wave pattern is observed at the same frequency. Determine the new wave speed down the string.

....

....

....

The length of the string will now be $\frac{3\lambda}{2}$.

(2 marks)

3 The figure below shows a standing wave on string:

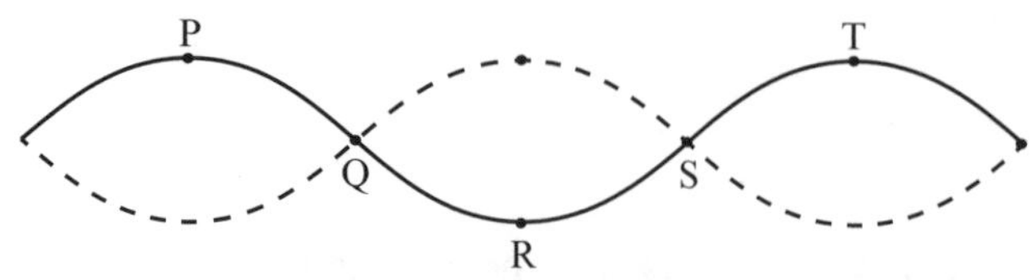

Which of the following pairs of points are moving in phase with each other?

☐ **A** P and R ☐ **B** Q and S ☐ **C** R and T ☐ **D** P and T **(1 mark)**

Stationary sound waves

1 The apparatus in the figure below can be used to determine the speed of sound.

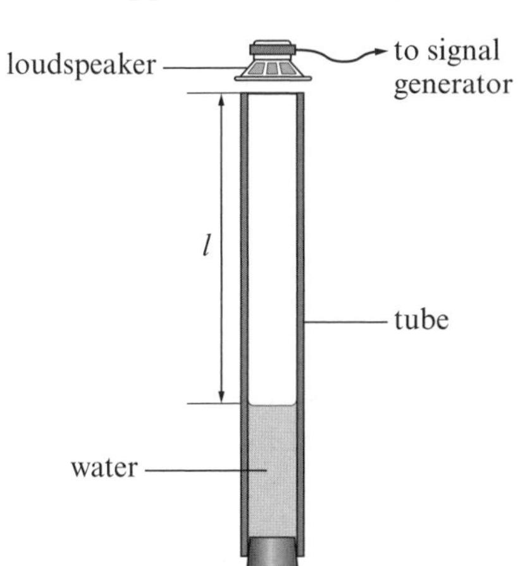

l / cm	f / Hz	f^{-1} / Hz^{-1}
9.3	920	0.00109
15.0	560	0.00179
22.7	370	0.00270
32.4	260	0.00385
37.5	220	0.00455
44.5	190	
50.7	170	

Calculations like this should be done to the same precision as the original data so that means to three significant figures including any zeros.

The method involves determining the lowest frequency, f, fed to the loudspeaker from the signal generator that will cause the air in the tube to resonate. Resonance is detected by listening carefully for loudest sound and then recording the frequency. The length of the air column in the tube, l, can be controlled by adding or removing water. The data obtained in the experiment are given in the table above.

(a) The air column is assumed to be vibrating in its fundamental mode. Explain the meaning of **fundamental mode**.

....

.... **(2 marks)**

(b) State the relationship between the length of the air column, l, and the wavelength of the sound, λ.

.... **(1 mark)**

(c) Complete the two blank cells in the data table at the top of this page. **(1 mark)**

(d) Use the grid to plot a graph of l against f^{-1}. **(3 marks)**

l / cm: 0, 10, 20, 30, 40, 50, 60

f^{-1} / Hz^{-1}: 0.000, 0.001, 0.002, 0.003, 0.004, 0.005, 0.006

(e) Determine the gradient of the graph and the speed of sound in the tube.

....

....

.... **(3 marks)**

Had a go ☐ Nearly there ☐ Nailed it! ☐

Stationary waves in closed and open tubes

1 The figure below represents a tube that is open at one end and closed at the other.

(a) Complete the diagram to show how the amplitude of the vibration of air in the tube varies along the length of the tube when it is vibrating at **three times its fundamental frequency**. **(2 marks)**

(b) Indicate the location of any displacement nodes with an **N** and any displacement antinodes with an **A**. **(2 marks)**

(c) Indicate the location of a pressure antinode with a **P**. **(1 mark)**

(d) Given that the internal length of the tube is 0.60 m and that the speed of sound in the tube is $340\,\mathrm{m\,s^{-1}}$, determine the frequency of vibration of the air in the pipe.

....

....

.... **(2 marks)**

2 A pipe is open at one end and closed at the other. It is found to resonate at 120 Hz. Which of the following is also a resonant frequency of the pipe?

> Open at one end and closed at the other means only odd harmonics are possible. From the options below, 120 Hz must be the fifth harmonic so f_0 is 24 Hz.

☐ **A** 48 Hz ☐ **B** 60 Hz ☐ **C** 72 Hz ☐ **D** 96 Hz **(1 mark)**

3 A tube is closed at one end and open at the other end. Its fundamental frequency is 80 Hz. If the same tube were open at both ends, its fundamental frequency would be:

☐ **A** 40 Hz ☐ **B** 80 Hz ☐ **C** 120 Hz ☐ **D** 160 Hz **(1 mark)**

4 The speed of sound in a gas is proportional to $\frac{1}{\sqrt{M}}$ where M is the molecular mass of the gas. Nitrogen has a molecular mass of 14 and that of helium is 4. The fundamental frequency of a pipe filled with nitrogen is 100 Hz. The same pipe filled with helium would have a fundamental frequency equal to:

☐ **A** 28.6 Hz ☐ **C** 187 Hz

☐ **B** 53.5 Hz ☐ **D** 350 Hz **(1 mark)**

> If $v \propto \frac{1}{\sqrt{M}}$ then $v\sqrt{M}$ is constant as is $f\sqrt{M}$ as $f \propto v$ so $f_N\sqrt{M_N} = f_H\sqrt{M_H}$ and $f_H = f_N\sqrt{\frac{M_N}{M_H}}$

Exam skills

1 The figure shows a ray of light entering a glass block with refractive index 1.50.

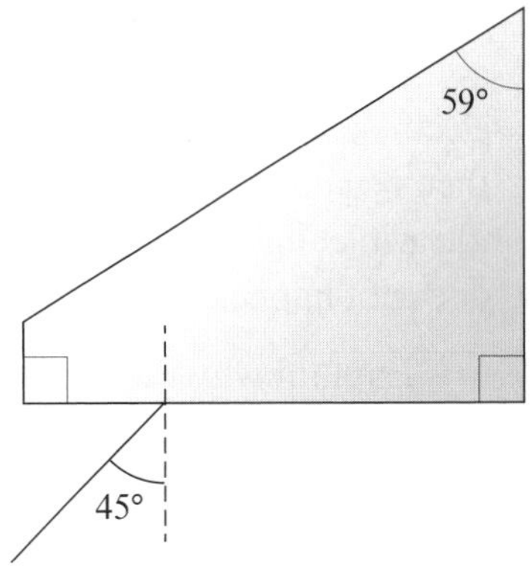

59°

45°

(a) Determine the angle of refraction as the ray enters the block.

....

....

....

.... **(2 marks)**

(b) Explain the term **critical angle** and how it relates to the phenomenon of **total internal reflection**. You should use a diagram to help with your answer.

....

....

....

.... **(3 marks)**

(c) Determine the critical angle for the glass.

....

....

.... **(2 marks)**

(d) Determine what happens to the ray when it strikes the sloping face of the block.

....

....

.... **(2 marks)**

(e) Sketch the path that the ray takes until it leaves the block on the figure at the top of this page. **(2 marks)**

Had a go ☐ Nearly there ☐ Nailed it! ☐

The photoelectric effect

1 An experiment to demonstrate the photoelectric effect uses a gold-leaf electroscope like the one shown here. The gold-leaf electroscope has a polished zinc plate placed on its cap, which is then negatively charged. When the zinc plate is illuminated with ultraviolet light, the gold-leaf electroscope is discharged. This does not happen when visible light of any intensity is used.

zinc plate

gold-leaf electroscope

(a) Explain why the experiment only works if the gold-leaf electroscope is initially charged negatively.

....

....

.... **(2 marks)**

(b) Explain why the gold-leaf electroscope is **not** discharged when illuminated with visible light.

....

....

.... **(2 marks)**

(c) Explain why increasing the intensity of the visible light does **not** have any effect on the gold-leaf electroscope.

....

....

.... **(2 marks)**

(d) Explain why the gold-leaf electroscope is discharged when illuminated with ultraviolet light.

....

....

.... **(2 marks)**

Guided

(e) What effect, if any, would be **observed** if ultraviolet light of greater intensity were used?

More UV photons per second means more photoelectrons are emitted per second ....

.... **(1 mark)**

(f) Explain why when a metal other than zinc is used on the top plate of the electroscope the threshold frequency changes.

....

....

.... **(2 marks)**

Einstein's photoelectric equation

1 The wavelength of yellow light from a particular source is 589 nm.

(a) Determine the energy, in joules, of a photon of the yellow light.

....

....

.... **(2 marks)**

Guided (b) What would the above photon energy be if expressed in electronvolts?

Joules are converted to electronvolts by dividing by 1.60×10^{-19}

.... **(1 mark)**

2 Define the work function, ϕ, of a metal.

....

.... **(1 mark)**

Guided **3** The work function of sodium is 2.36 eV.

(a) Determine the threshold frequency of sodium.

The work function $\phi = hf_0$ so $f_0 = \frac{\phi}{h} =$ ....

.... **(2 marks)**

(b) Determine the maximum kinetic energy of a photoelectron emitted from the surface of a piece of sodium that is illuminated by light of wavelength 465 nm.

....

.... **(2 marks)**

4 The work function of potassium is 2.29 eV. Show that visible light could result in the emission of photoelectrons from a piece of potassium.

....

.... **(2 marks)**

5 The graph shows how the maximum kinetic energy of photoelectrons varies with the frequency of incident light on a metal surface.

Use the graph to determine the work function of the metal.

....

....

....

(2 marks)

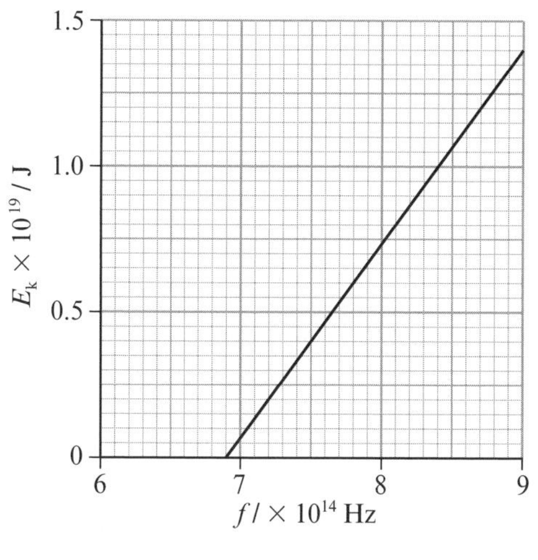

Had a go ☐ Nearly there ☐ Nailed it! ☐

Determining the Planck constant

1 A method of determining the Planck constant involves using a number of differently coloured light-emitting diodes with different peak wavelengths λ and observing the forward voltage V_F across them just as they switch on.

The circuit shown below can be used. Results are shown in the table.

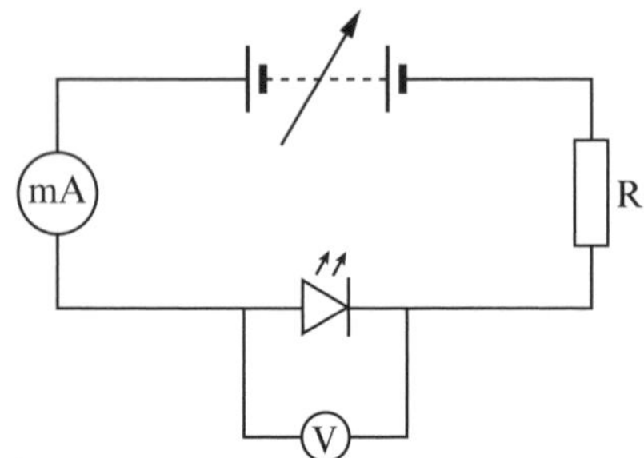

Colour	λ/nm	λ^{-1}/m^{-1} × 10^6	V_F/V
red	635	1.57	1.96
orange	612	1.63	2.03
yellow	585	1.71	2.12
green	555	1.80	2.24
blue-green	505	1.98	2.46
blue	430	2.33	2.89

(a) Suggest a reason for the inclusion of the resistor, R, in the circuit.

....

.... **(1 mark)**

(b) Determine the energy transferred as an electron passes through a red LED.

....

.... **(2 marks)**

Guided (c) Show that the forward voltage drop across the LEDs is given by $V_F = \frac{hc}{e\lambda}$

Photon energy = energy transferred, so $hf = eV_F$, and as $c = f\lambda$

....

....

.... **(2 marks)**

(d) Using the blank grid, plot a graph of V_F against wavelength^{-1}.

(5 marks)

(e) Use the graph to determine the Planck constant.

....

....

.... **(2 marks)**

Electron diffraction

1 Electrons are accelerated by a potential difference of 3.00 kV in an electron gun before striking a graphite target and subsequently forming a pattern of diffraction rings on a phosphor screen (see figure below).

vacuum tube
diffraction rings
electron gun
graphite target
phosphor screen

(a) Explain how this experiment provides evidence for the wave nature of electrons.

....

.... **(2 marks)**

Guided

(b) The target is polycrystalline graphite. Explain why this results in diffraction rings.

Diffraction occurs because the wavelength of the electrons is comparable to the atomic spacing in the graphite. 'Polycrystalline' means made of very many crystals. These will have countless different orientations, leading to ....

.... **(2 marks)**

(c) Electrons have mass m and momentum p. Their kinetic energy is equal to:

☐ **A** $\frac{p}{2m}$ ☐ **B** $\sqrt{\frac{p}{2m}}$ ☐ **C** $\frac{p^2}{m}$ ☐ **D** $\frac{p^2}{2m}$ **(1 mark)**

Guided

(d) Show that electrons leave the electron gun at about $3 \times 10^7\,\mathrm{m\,s^{-1}}$.

The electrons of charge e are accelerated by a p.d. V and gain kinetic energy equal to eV. This means that $eV = \frac{1}{2}mv^2$ and $v = \sqrt{\frac{2eV}{m}}$

....

.... **(2 marks)**

Guided

(e) Determine the de Broglie wavelength of the electrons.

The de Broglie wavelength of a particle is given by $\lambda = \frac{h}{p}$ where p is the momentum, mv, and h is the Planck constant, which is equal to 6.63×10^{-34} J s. ....

.... **(1 mark)**

(f) The accelerating voltage is increased. The rings will now be:

☐ **A** of smaller radius and brighter ☐ **C** of larger radius and brighter

☐ **B** of smaller radius and less bright ☐ **D** of larger radius and less bright. **(1 mark)**

Had a go ☐ Nearly there ☐ Nailed it! ☐

Wave–particle duality

1 (a) Name an experiment that provides evidence for the wave nature of light and explain how it does so.

....

Think of the properties of waves – for example, diffraction, refraction, interference – and a demonstration that light has these properties.

....

....

....

(3 marks)

(b) Name an experiment that provides evidence for the particle nature of light and explain how it does so.

....

Think of the properties of particles – for example, their discrete nature – and a demonstration that light has these properties.

....

....

....

(3 marks)

2 When light of a very low intensity is shone through a pair of slits, individual photons can be observed with a suitable detector where one might ordinarily see interference fringes. More photons are observed where one would see bright fringes, leading to the idea that the intensity of light is related to the likelihood of finding photons.

(simulation)

Explain how this experiment confirms ideas of wave–particle duality.

....

....

....

....

....

....

(3 marks)

Exam skills

1 These questions concern the photoelectric effect and Einstein's equation.

(a) Explain the meaning of the term **threshold frequency** in the context of the photoelectric effect.

....

....

.... **(2 marks)**

The sketch graph below shows how the kinetic energy of photoelectrons varies with the frequency of incident light on a metal surface.

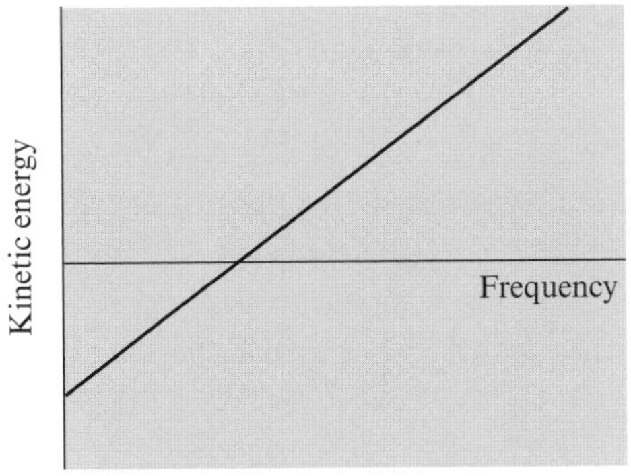

Kinetic energy

Frequency

(b) Label the threshold frequency, f_0. **(1 mark)**

(c) Explain how the threshold frequency is related to the work function of the metal used.

....

....

.... **(2 marks)**

(d) Add a second line to the graph that represents the behaviour of a metal with a higher work function. **(2 marks)**

Einstein's equation states that the energy of incident photons is equal to the work function added to the maximum kinetic energy of emitted photoelectrons, or $hf = \phi + KE_{max}$

(e) Describe the nature of a photon.

....

....

.... **(2 marks)**

(f) Explain why the final term in the equation is the **maximum** kinetic energy of emitted photoelectrons.

....

....

.... **(2 marks)**

Had a go ☐ Nearly there ☐ Nailed it! ☐

Temperature and thermal equilibrium

1 Which one of the following statements about temperature is incorrect?

☐ **A** A temperature rise of 10 °C is the same as a temperature rise of 10 K.

☐ **B** The temperature of snow is around 273 K.

☐ **C** The boiling point of nitrogen (−196 °C) is 77 K.

☐ **D** The melting point of lead (3422 °C) is 3149 K. **(1 mark)**

2 Explain what is meant by 'Brownian motion'.

....

....

.... **(2 marks)**

Guided

3 Explain what is meant by:

(a) thermal equilibrium

When two bodies are in thermal equilibrium ....

....

(b) internal energy

Internal energy is the sum of ....

....

(c) temperature.

Temperature is a measure of ....

....

It is related to the amount of kinetic energy ....

.... **(4 marks)**

4 (a) Convert the temperatures below from °C to K:

(i) boiling point of bromine: 58.8 °C

....

(ii) boiling point of oxygen: −183 °C

....

(b) The temperature at the surface of Venus is 735 K. Convert this temperature from K to °C.

.... **(3 marks)**

Solids, liquids and gases

1 The diagram shows how potential energy varies with molecular separation inside a solid material. Which position, A, B, C or D, represents the equilibrium bond length in the material?

.... **(1 mark)**

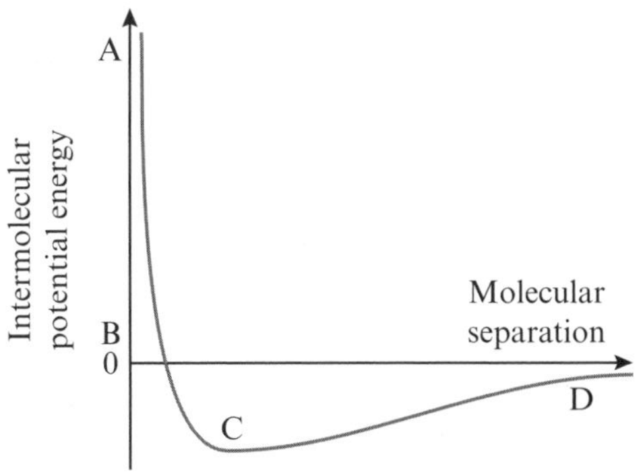

Guided

2 The graph shows how the temperature of a fixed mass of water varies as it is heated. The energy is supplied at a constant rate.

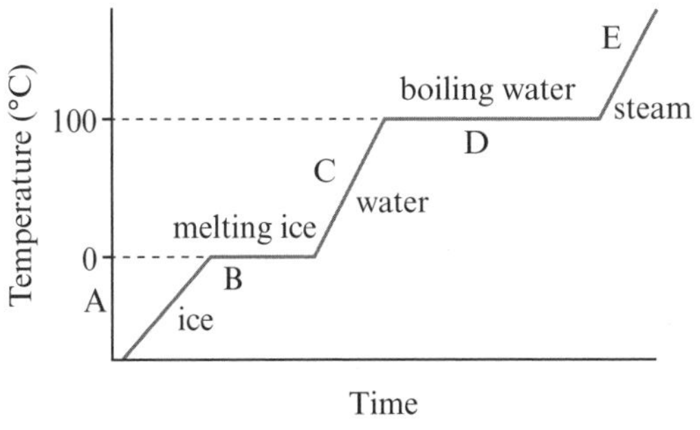

(a) Explain why the temperature rises in sections A, C and E but not in sections B and D.

In sections A, C and E there is no change of

....

All the thermal energy supplied increases the

....

of the molecules, causing a ....

In sections B and D there is a change of ....,

so the thermal energy supplied is increasing the ....

of the molecules as it breaks ....,

with no rise in .... **(4 marks)**

(b) What is represented by the region of the graph in section B?

.... **(1 mark)**

(c) What is represented by the region of the graph in section C?

....

.... **(1 mark)**

3 When a volatile liquid evaporates it has a cooling effect.

(a) Explain, in molecular terms, what happens when a liquid evaporates.

....

.... **(2 marks)**

(b) Explain why evaporation has a cooling effect.

....

.... **(2 marks)**

Had a go ☐ Nearly there ☐ Nailed it! ☐

Specific heat capacity

1 A block of ice of mass 2.5 kg is at a temperature of −15 °C. How much energy must be supplied to raise the temperature of the ice to its melting point? The specific heat capacity of the ice is $2100\,\mathrm{J\,kg^{-1}\,°C^{-1}}$.

....

.... **(2 marks)**

2 Describe an experiment to measure the specific heat capacity of a non-flammable liquid such as water. Include a labelled diagram and explain how you would use the measurements taken to calculate a value for the specific heat capacity. Explain any precautions you would take to ensure an accurate result.

Make sure your diagram is fully labelled!

....

....

....

Practical skills Think of ways to reduce the effect of heat losses in order to ensure a more accurate result.

....

....

....

....

.... **(8 marks)**

Guided

3 A block of copper of mass 250 g and at a temperature of 50 °C is submerged in $0.0020\,\mathrm{m^3}$ of water at a temperature of 20 °C. Calculate the final temperature of the water and the block, assuming there are no heat losses to the surroundings. (Specific heat capacity of copper = $385\,\mathrm{J\,kg^{-1}\,°C^{-1}}$; specific heat capacity of water = $4200\,\mathrm{J\,kg^{-1}\,°C^{-1}}$; density of water = $1000\,\mathrm{kg\,m^{-3}}$.)

Equate the heat lost by the copper to the heat gained by the water.

Energy transferred from copper to water $E = m_{Cu}c_{Cu}\Delta\theta_{Cu} = m_W c_W \Delta\theta_W$

....

....

if final temperature of water and copper = T, then $\Delta\theta_{Cu} = 50 - T$ and

$\Delta\theta_W = T - 20$ ....

....

.... **(4 marks)**

Specific latent heat 1

1 The diagrams show stages in the 'method of mixtures' used to determine the latent heat of fusion of ice.

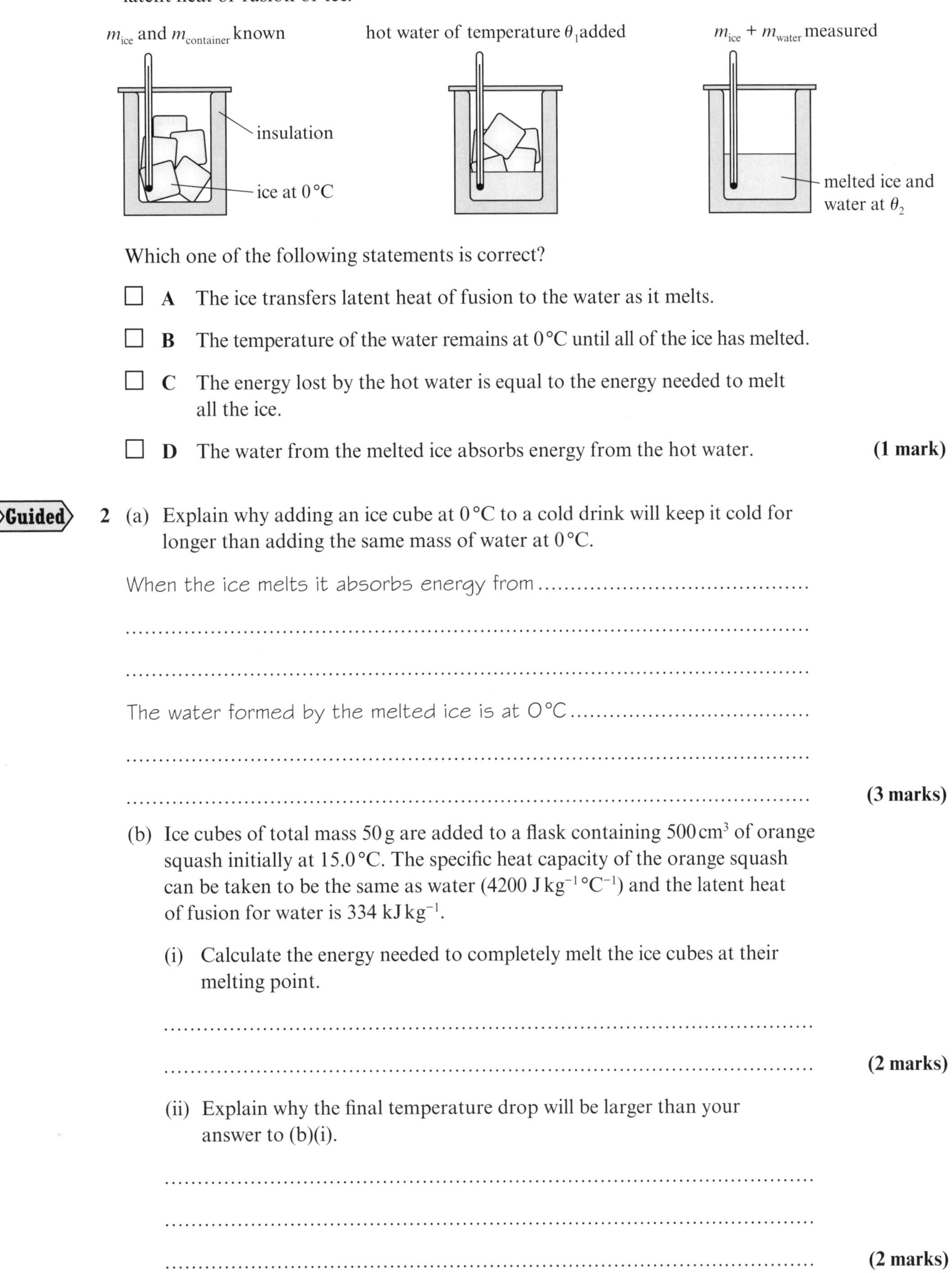

Which one of the following statements is correct?

☐ **A** The ice transfers latent heat of fusion to the water as it melts.

☐ **B** The temperature of the water remains at 0 °C until all of the ice has melted.

☐ **C** The energy lost by the hot water is equal to the energy needed to melt all the ice.

☐ **D** The water from the melted ice absorbs energy from the hot water. **(1 mark)**

Guided

2 (a) Explain why adding an ice cube at 0 °C to a cold drink will keep it cold for longer than adding the same mass of water at 0 °C.

When the ice melts it absorbs energy from ....

....

....

The water formed by the melted ice is at 0 °C ....

....

.... **(3 marks)**

(b) Ice cubes of total mass 50 g are added to a flask containing 500 cm^3 of orange squash initially at 15.0 °C. The specific heat capacity of the orange squash can be taken to be the same as water (4200 J kg^{-1} °C^{-1}) and the latent heat of fusion for water is 334 kJ kg^{-1}.

(i) Calculate the energy needed to completely melt the ice cubes at their melting point.

....

.... **(2 marks)**

(ii) Explain why the final temperature drop will be larger than your answer to (b)(i).

....

....

.... **(2 marks)**

Specific latent heat 2

1 Burns from steam can be more serious than burns from boiling water. Which of the following statements gives the best explanation for this?

☐ **A** Steam always has a much higher temperature than boiling water.

☐ **B** Boiling water has a low specific heat capacity.

☐ **C** When steam condenses to water it releases a lot of energy as bonds form.

☐ **D** Steam is a much better conductor of heat than boiling water. **(1 mark)**

Guided

2 Calculate the thermal energy that must be supplied to change 1.20 kg of water at 25.0 °C to steam at 100 °C. The specific heat capacity of water is 4200 $J\,kg^{-1}\,K^{-1}$ and the specific latent heat of vaporisation of water is 2.26 $MJ\,kg^{-1}$.

The energy needed to raise the temperature from 25 °C to 100 °C

is $E = mc\Delta\theta =$ ....

The additional energy needed to change it from water to steam at

100 °C is ....

.... **(3 marks)**

3 An early method for finding the specific latent heat of vaporisation of a liquid used an apparatus like the one shown here.

Practical skills

(a) Explain how this apparatus was used to find the latent heat of vaporisation of a liquid. Your answer should include a list of the measurements to be made and should explain how the results are used to calculate a value for the latent heat of vaporisation.

....

....

....

....

....

....

.... **(6 marks)**

Start by explaining how the apparatus works: electrical energy supplied to the heater, ... Then state the measurements needed: mass, ... Finally, describe how to calculate latent heat.

(b) Explain why it would be important to wait until the liquid is boiling before beginning to measure the electrical energy supplied to the heater.

....

.... **(2 marks)**

Kinetic theory of gases

1 Which one of the following statements is **not** an assumption of the kinetic theory model of a gas?

☐ **A** The volume of the molecules is small compared with the volume of their container.

☐ **B** Molecular collisions are all perfectly elastic.

☐ **C** Gases obey the ideal gas equation.

☐ **D** There are no long range forces between the molecules. **(1 mark)**

2 Use the kinetic theory model to explain why:

(a) Gases exert a pressure on the walls of their containers.

....

....

.... **(3 marks)**

(b) The pressure of a gas increases when it is compressed at constant temperature.

....

....

.... **(3 marks)**

3 When a volatile liquid evaporates, the molecules diffuse through the air but the rate of diffusion is much slower than the mean speed of the molecules.

(a) Use the kinetic theory to explain why the rate of diffusion is so slow.

....

....

.... **(2 marks)**

Guided

(b) Suggest, with reasons, how the rate of diffusion will depend on the molecular mass and temperature.

The rate of diffusion will depend on the rms speed of the molecules.

From $pV = \frac{1}{3}Nmc_{rms}$ we can see that $\sqrt{c_{rms}}$ is proportional ....

We also know that $\frac{1}{2}mc_{rms} = \frac{3}{2}kT$ so ....

....

....

.... **(4 marks)**

Maths skills The rate of diffusion is likely to depend on the speed of the molecules, so use the kinetic theory equations to see how the rms speed depends on *m* and *T*.

The gas laws: Boyle's law

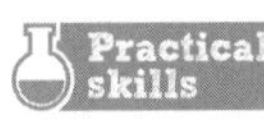

1 The diagram shows some apparatus that can be used to measure how the volume of a gas depends on its pressure. Pressure is applied to the oil in the reservoir, forcing the oil into the tube and, in turn, compressing the air trapped in the tube. The length of the air column in the tube can be measured against the scale.

(a) Explain why it is important for the air column to be trapped in a tube of constant cross-sectional area.

....

.... **(2 marks)**

(b) Explain why it is important to compress the gas **slowly**.

....

.... **(2 marks)**

(c) Boyle's law suggests that the volume of a constant mass of an ideal gas at constant temperature is inversely proportional to the pressure of the gas. Explain how results taken from an experiment like the one shown above can be used to test this relationship.

> What graph could be plotted? Simply plotting volume (or length of air column) against pressure will not be good enough.

....

....

....

....

.... **(4 marks)**

Guided

2 (a) The pressure inside an oxygen cylinder is 1.2×10^7 Pa. The cylinder has a volume of $0.0040\,\text{m}^3$. The oxygen is released into the atmosphere. Calculate the volume occupied by the oxygen from the cylinder when it is at atmospheric pressure (1.0×10^5 Pa).

Using Boyle's law: $p_1V_1 = p_2V_2$ (where 1 refers to the initial state and 2 to the final state) ....

....

.... **(2 marks)**

(b) State **two** assumptions that you made in carrying out your calculation in part (a) above.

....

....

.... **(2 marks)**

The gas laws: the pressure law

Practical skills

1 (a) The diagram shows an experimental set-up that can be used to test the pressure law. Explain how you could use this apparatus to find a value in degrees Celsius for the absolute zero of temperature.

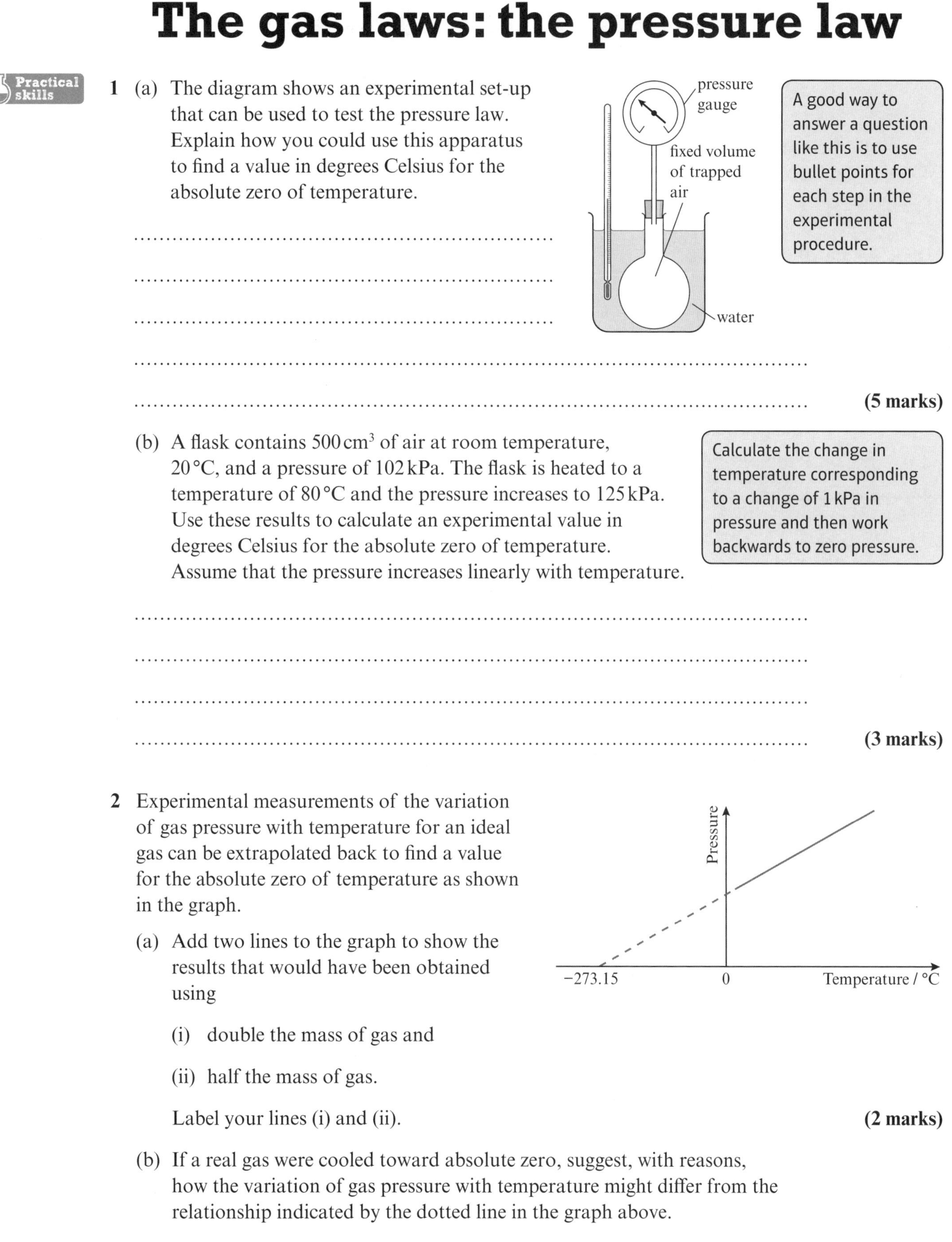

A good way to answer a question like this is to use bullet points for each step in the experimental procedure.

....

....

....

....

.... **(5 marks)**

(b) A flask contains 500 cm^3 of air at room temperature, 20 °C, and a pressure of 102 kPa. The flask is heated to a temperature of 80 °C and the pressure increases to 125 kPa. Use these results to calculate an experimental value in degrees Celsius for the absolute zero of temperature. Assume that the pressure increases linearly with temperature.

Calculate the change in temperature corresponding to a change of 1 kPa in pressure and then work backwards to zero pressure.

....

....

....

.... **(3 marks)**

2 Experimental measurements of the variation of gas pressure with temperature for an ideal gas can be extrapolated back to find a value for the absolute zero of temperature as shown in the graph.

(a) Add two lines to the graph to show the results that would have been obtained using

(i) double the mass of gas and

(ii) half the mass of gas.

Label your lines (i) and (ii). **(2 marks)**

(b) If a real gas were cooled toward absolute zero, suggest, with reasons, how the variation of gas pressure with temperature might differ from the relationship indicated by the dotted line in the graph above.

....

....

....

.... **(3 marks)**

Had a go ☐ Nearly there ☐ Nailed it! ☐

The equation of state of an ideal gas

1 A constant mass of an ideal gas is compressed slowly so that its temperature does not change. Which of the following quantities remains the same?

☐ **A** pressure

☐ **B** volume

☐ **C** pressure × volume

☐ **D** pressure ÷ volume

(1 mark)

2 The graph shows how the volume of a constant mass of an ideal gas at constant pressure varies with temperature in Celsius.

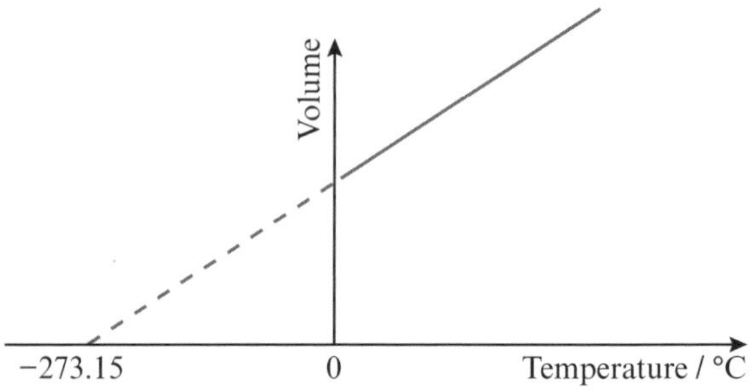

(a) Add a line to the graph to show how the results would differ if the pressure of the gas was greater. Label this line X.

(2 marks)

Volume

−273.15

0

Temperature / °C

(b) Add another line to the graph to show how the results would differ if half the mass of the original gas had been used. Label this line Y. **(2 marks)**

3 Avogadro's law states that equal volumes of all gases at the same temperature and pressure contain the same number of molecules.

(a) Show that Avogadro's law is a consequence of the equation of state for an ideal gas.

....

....

....

.... **(3 marks)**

Guided

(b) When oxygen and hydrogen react together to form water, the ratio of the volumes of gases involved in the reaction is 2:1 (measured under the same conditions of temperature and pressure). Explain how this provides evidence that water molecules contain two hydrogen atoms and one oxygen atom.

By Avogadro's law there are twice as many .... as .... involved in the reaction (twice the volume of ....).

The simplest way to combine the .... is 2 ....

to 1 ....(H_2O). **(3 marks)**

The kinetic theory equation

1 The total internal energy of an ideal gas is doubled. How does this affect the root mean square (rms) speed of the molecules in the gas?

☐ **A** rms speed is not affected

☐ **B** rms speed increases by a factor of $\sqrt{2}$

☐ **C** rms speed increases by a factor of 2

☐ **D** rms speed increases by a factor of 4.

(1 mark)

2 Six molecules in a gas have speeds: 406, 438, 395, 450, 448, 508 m s^{-1}.

(a) Calculate the mean speed c_{rms}.

....

(b) Calculate the mean square speed c_{rms}.

....

(c) Calculate the root mean square speed c_{rms}.

.... **(3 marks)**

3 The graph shows the distribution of particle energies in a gas at two different temperatures, T_1 and T_2.

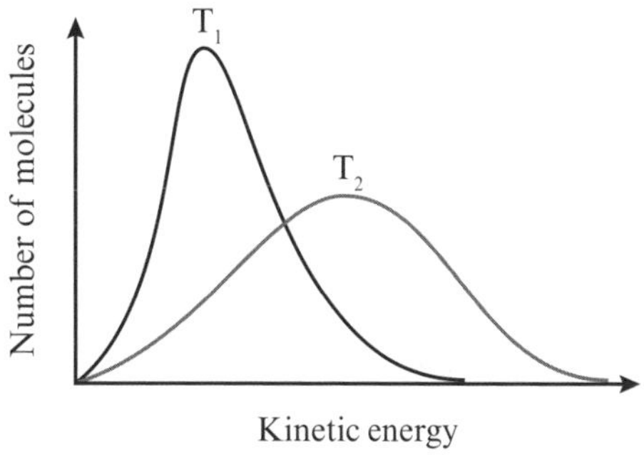

State, with reasons, which of T_1 and T_2 is the higher temperature.

....

....

.... **(2 marks)**

4 A hot air balloon contains air at a temperature of 40 °C.

(a) Calculate the mean kinetic energy of molecules in this balloon.

....

.... **(2 marks)**

(b) Calculate the root mean square speed of an air molecule in this balloon (take the mass of an air molecule to be approximately 4.7×10^{-26} kg).

....

.... **(2 marks)**

(c) Air consists of a mixture of gases of different molecular mass. Explain why the molecular mass does not affect the mean kinetic energy of the molecules of each gas in the balloon.

....

.... **(2 marks)**

The internal energy of a gas

1 When carbon dioxide is released from a pressurised cylinder into the atmosphere there is a sudden and significant drop in temperature. This cooling effect can be used to form dry ice (solid carbon dioxide). Which of the statements below gives a correct explanation for the sudden drop in temperature?

☐ **A** The carbon dioxide in the cylinder was much colder than the atmosphere.

☐ **B** The carbon dioxide expands rapidly and its internal energy falls.

☐ **C** The carbon dioxide molecules are more massive than air molecules.

☐ **D** The atmosphere does work on the expanding carbon dioxide. **(1 mark)**

2 The air in a child's balloon is at a temperature of 25 °C and a pressure of 110 kPa. The volume of the balloon is $3.0 \times 10^{-3}\,m^3$.

(a) Calculate the number of moles of air in the balloon and hence the number of molecules inside the balloon.

....

....

....

.... **(4 marks)**

Guided

(b) Calculate the total internal energy of the gas in the balloon. (Assume that the internal energy of the gas is equal to the total kinetic energy of its molecules).

The internal energy of n moles of an ideal gas is given by $U = \frac{3}{2}nRT$

....

....

.... **(2 marks)**

3 On a hot day the temperature inside a room is 30 °C.

(a) Calculate the mean kinetic energy of an air molecule in this room.

Don't forget to convert to kelvin!

....

....

.... **(2 marks)**

(b) Calculate the root mean square speed of an air molecule in this room (take the mass of an air molecule to be approximately 4.7×10^{-26} kg).

....

....

....

.... **(2 marks)**

Exam skills

1 Describe an experiment that can be used to investigate how the pressure of a constant volume and mass of gas varies with temperature. Your answer should include (i) a labelled diagram of the apparatus, (ii) an explanation of how the experiment is carried out and what measurements must be taken.

....

....

....

....

.... **(5 marks)**

2 Explain how the results of such an experiment can be used to find a value in degrees Celsius for the absolute zero of temperature.

....

....

.... **(2 marks)**

3 How is the existence of an absolute zero of temperature explained in terms of the kinetic theory?

....

....

.... **(2 marks)**

4 A small amount of bromine gas (Br_2) escapes into the air in a laboratory. The air temperature is 22 °C. The mass of a bromine molecule is 2.7×10^{-25} kg.

(a) How does the mean kinetic energy of the bromine molecules compare with the mean kinetic energy of the air molecules? (Do **not** carry out a calculation.)

....

.... **(2 marks)**

(b) Calculate the rms speed of these bromine molecules in air.

....

....

.... **(3 marks)**

Had a go ☐ Nearly there ☐ Nailed it! ☐

Angular velocity

1 The drum of a washing machine is rotating at constant angular velocity ω. Which of the following statements about the motion of a point on the outside of the rotating drum is correct?

☐ **A** The time for it to complete one rotation is $2\pi\omega$.

☐ **B** Its velocity is constant.

☐ **C** It is accelerating.

☐ **D** There is a resultant outward force acting on it. **(1 mark)**

2 The Earth orbits the Sun in one year at a mean distance of 150 million kilometres. Assume the orbit is circular. There are 3.16×10^7 s in one year.

(a) Calculate the angular velocity of the Earth as it orbits the Sun. Give your answer in radians per second.

Make sure you convert km to m and years to seconds.

....

....

.... **(3 marks)**

(b) Calculate the Earth's orbital speed in m s^{-1}.

....

.... **(1 mark)**

(c) Calculate the tangential speed of a man standing on the Earth's equator as the Earth rotates on its axis in a day. The radius of the Earth is 6400 km.

....

....

.... **(2 marks)**

Guided

(d) The time taken for Mars to complete one orbit of the Sun is 1.87 Earth years. At a particular time, Earth (E) and Mars (M) lie along the same line from the Sun (S). After one Earth year they are at different points in their orbit as shown in the diagram. Assuming a circular orbit for Mars, calculate the angle θ in radians.

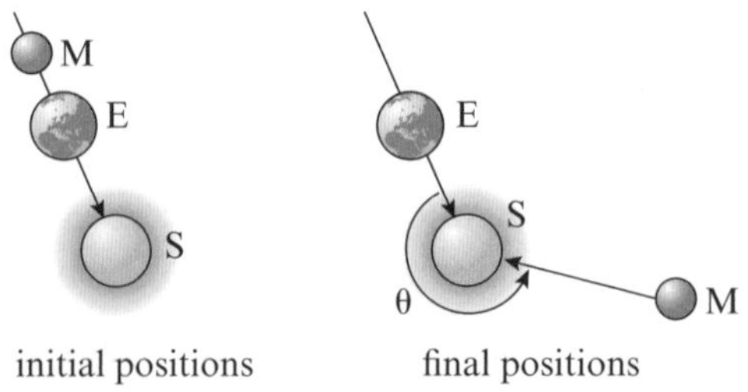

After one year the Earth will be back in its original position so the angle θ is just the angular displacement of Mars during one Earth year.

$\Delta T = 1$ earth year $= 3.16 \times 10^7$ s.

Angular displacement $\theta = \omega \Delta T$

$$\omega = \frac{2\pi}{T} = \frac{2\pi}{(1.87 \times 3.16 \times 10^7)}$$

....

.... **(3 marks)**

Centripetal force and acceleration

1 A racing car of mass 1200 kg is cornering at constant speed of 18 m s^{-1} and follows a path that is part of a horizontal circle of radius 45 m.

(a) Calculate the magnitude of the resultant force acting on the car and state its direction.

....

....

.... **(2 marks)**

(b) Explain how this resultant force is produced.

....

....

....

.... **(2 marks)**

(c) On the next lap, the driver of the car attempts to take the corner at a higher speed. Explain why the car might skid off the track.

> There is no outward force acting on the car, but the inward force is limited.

....

....

.... **(2 marks)**

2 A simple pendulum consists of a light, inextensible string of length 0.50 m and a small bob of mass 0.060 kg. The bob is displaced sideways so that it rises through a vertical height of 0.050 m and is then released.

Guided

(a) Explain why the tension in the string at the moment the bob passes through its lowest position is greater than the weight of the bob.

When the bob moves through its lowest position, it is moving in circular motion, so there must be a resultant force towards ....

....

.... **(3 marks)**

(b) Calculate the tension in the string at the moment the bob passes through its lowest position.

> Sketch a free-body diagram of the bob as it passes the lowest position and use this to find an expression for the tension force.

....

....

....

.... **(4 marks)**

Simple harmonic motion

Practical skills

1 A student wants to measure the frequency of a simple pendulum. Describe a suitable method explaining how to ensure an accurate result.

The key word here is 'accurate'. How will you minimise the significance of timing errors? How will you make sure you count from the same position on each swing? How will you use repeats and averages?

....

....

....

....

....

....

.... **(5 marks)**

2 When a ruler is fixed to a bench so that part of it extends beyond the edge of the bench, the extended part will undergo vertical oscillations when displaced and released. A student is trying to find out whether the oscillations are simple harmonic. In order to do this, she attaches a mass to the end of the ruler and measures its vertical displacement. She repeats the procedure for a range of masses.

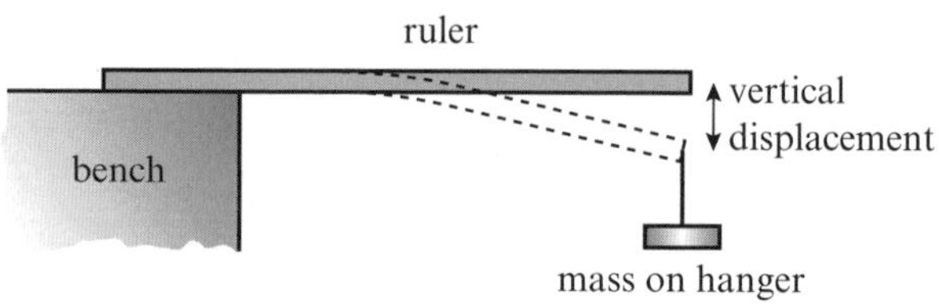

ruler
bench
vertical displacement
mass on hanger

She obtains the following data:

Mass / g	10	20	30	40	50	60	70
Deflection / mm	3.0	6.0	9.0	11.8	13.4	15.7	17.8

(a) State the **two** conditions necessary for an oscillator to undergo simple harmonic oscillations.

....

....

.... **(2 marks)**

(b) Discuss whether the ruler will undergo s.h.m. when displaced for the range of masses used. Consider the effect of small and large initial displacements of the end of the ruler.

Look at the data and see if it fits the criteria for s.h.m. in part (a).

....

....

....

....

....

.... **(4 marks)**

Solving the s.h.m. equation

1 A simple harmonic oscillator is released from an initial displacement of 10 cm at time $t = 0$. It then oscillates with a time period of 0.20 s. Which of the equations below gives its displacement as a function of time?

☐ **A** $10\cos(10\pi t)$ ☐ **C** $10\cos(0.40\pi t)$

☐ **B** $20\cos(10\pi t)$ ☐ **D** $20\cos(0.40\pi t)$ **(1 mark)**

Maths skills

2 A simple harmonic oscillator of mass 0.25 kg has an amplitude of 8.0 cm and oscillates with a frequency of 2.0 Hz. At time $t = 0$ it has a positive displacement of 8.0 cm.

(a) Calculate the time period of the oscillation.

.... **(1 mark)**

(b) Calculate the displacement of the oscillator after:

(i) 0.125 s

....

(ii) 0.25 s

....

(iii) 0.40 s

.... **(3 marks)**

(c) Calculate the maximum speed of the oscillator and state the value of the displacement at which this occurs.

....

.... **(2 marks)**

(d) Calculate the maximum force acting on the oscillator and state the displacements at which it occurs.

....

.... **(2 marks)**

Guided

(e) Calculate the total energy of the oscillator.

Maximum KE (at maximum velocity) = ...

At this point the potential energy is ....

.... **(2 marks)**

Maths skills

3 The displacement x of a simple harmonic oscillator is $x = A\cos(2\pi ft)$.

Write down equations for the velocity and acceleration of this oscillator.

....

.... **(2 marks)**

Had a go ☐ Nearly there ☐ Nailed it! ☐

Graphical treatment of s.h.m.

1 The diagram below shows how the displacement of a particular simple harmonic oscillator varies with time.

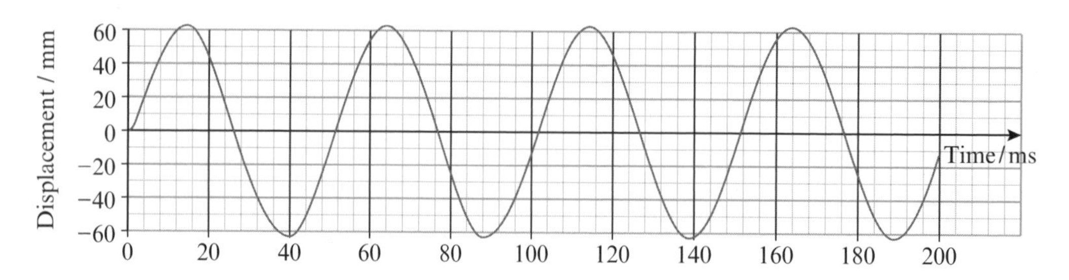

(a) State the amplitude of this oscillation.

.... **(1 mark)**

Maths skills

(b) Calculate the frequency of this oscillation.

.... **(1 mark)**

(c) Draw, using the axes below, a graph to show the variation of velocity with time for this oscillation. Add a suitable scale to the velocity axis with a period of about 52 ms.

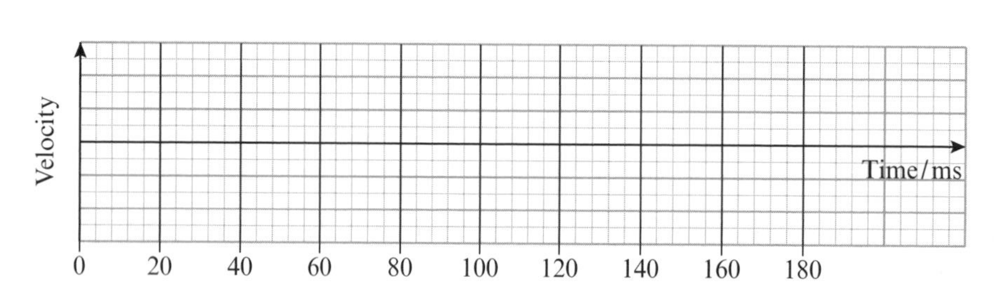

Velocity is the rate of change of displacement so it is related to the gradient of the displacement against time graph.

(4 marks)

(d) Draw, using the axes below, a graph to show the variation of acceleration with time for this oscillation. Add a suitable scale to the acceleration axis with a period of about 52 ms.

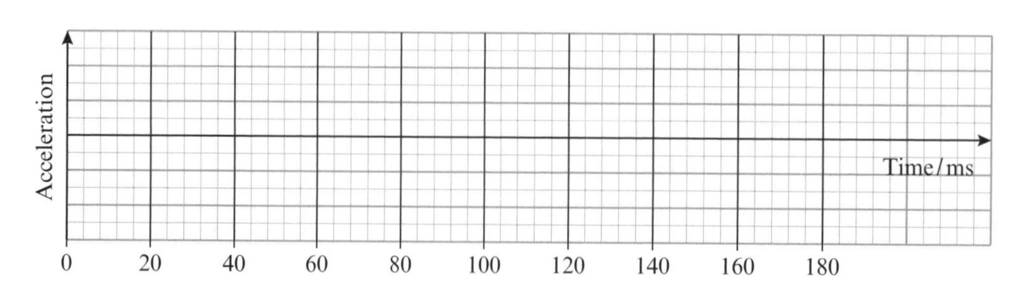

(4 marks)

Energy in s.h.m.

1 The frequency and amplitude of vibration of a loudspeaker are both doubled. By what factor does the energy of the oscillations change?

☐ **A** ×2 ☐ **B** ×4 ☐ **C** ×8 ☐ **D** ×16 **(1 mark)**

2 A mass of 500 g is suspended from a spring of spring constant $42\,\text{N}\,\text{m}^{-1}$. It is lifted 2.0 cm vertically and then released.

(a) What energy transfers take place during one complete oscillation?

....

.... **(2 marks)**

(b) Calculate the extension of the spring when the mass is at its equilibrium.

....

.... **(2 marks)**

(c) Calculate the energy stored in the spring when the mass is at:

The energy stored in a stretched spring is given by $E = \frac{1}{2}kx^2$.

(i) its equilibrium position

.... **(1 mark)**

(ii) 2.0 cm above its equilibrium position

.... **(1 mark)**

(iii) 2.0 cm below its equilibrium position.

.... **(1 mark)**

(d) Calculate the change in gravitational potential energy of the mass between the top and bottom of its first oscillation and hence show that the loss of gravitational energy is equal to the gain in elastic potential energy between these two positions.

....

.... **(2 marks)**

Guided

(e) Calculate the maximum kinetic energy of the oscillator and hence calculate its maximum velocity.

Maximum KE is equivalent to maximum GPE.

max. KE = ...

$KE = \frac{1}{2}mv^2$ so v = ...

.... **(3 marks)**

(f) The mass spring system is lightly damped and after 100 complete oscillations its amplitude has fallen to 0.50 cm.

What fraction of the original energy remains in the oscillator after 100 oscillations?

.... **(1 mark)**

Forced oscillations and resonance

1 A student carries out an experiment to investigate the response of a mass–spring oscillator to a driving force using the apparatus shown in the diagram.

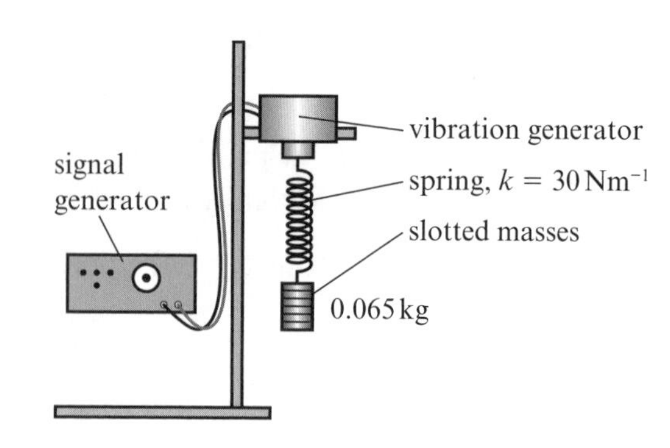

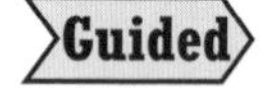

(a) (i) State what is meant by natural frequency and forcing frequency.

Natural frequency is the frequency of oscillations when the oscillator is displaced and released with no external forces acting on it.

The forcing frequency is ....

.... **(2 marks)**

(ii) Calculate the natural frequency of oscillations of this mass–spring oscillator.

....

.... **(1 mark)**

(b) (i) In the experiment, the student sets the signal generator to a small fixed amplitude A_0 and then gradually increases the frequency from 0 to 10 Hz. Use the axes to show the expected response of the oscillator. Assume that damping forces are small (but not zero).

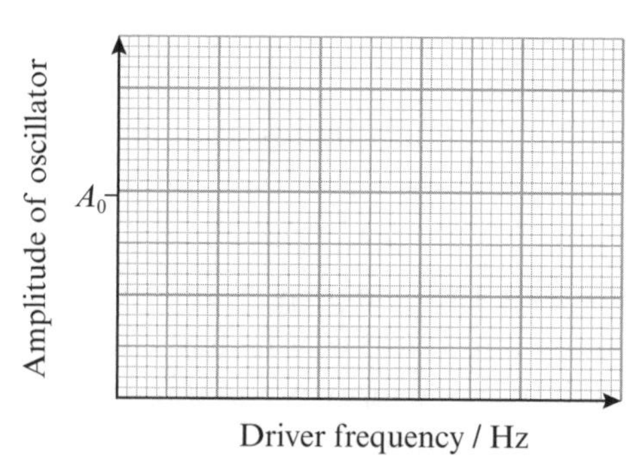

(3 marks)

(ii) Explain what is meant by resonance and state when it occurs.

....

.... **(2 marks)**

(iii) Describe the energy transfers taking place at resonance.

....

....

.... **(2 marks)**

(c) In a second experiment, the student sticks a horizontal cardboard disc to the bottom of the masses to act as a damper. Draw a second line on the graph above to show the effect of dampin. **(2 marks)**

Exam skills

1 Modern mountain bikes have suspension springs on their front forks, so that the front wheel can move up and down relative to the frame. The spring on a particular mountain bike has a spring constant of $6500\,\text{N m}^{-1}$ and it compresses by 0.084 m when the bike hits a bump. The mass of bike and rider is 75 kg.

(a) Assuming the bike and rider system together acts like a mass–spring oscillator, calculate:

(i) The natural frequency of simple harmonic oscillations after hitting the bump.

....

....

.... **(3 marks)**

(ii) The maximum energy stored in the spring.

....

.... **(2 marks)**

(iii) The maximum vertical acceleration of the bike during the oscillations.

....

.... **(2 marks)**

(iv) The rider has a mass of 60 kg. Calculate the maximum vertical reaction force from his saddle after he hits the bump.

....

....

.... **(2 marks)**

(b) (i) Explain why a damper is also included in the front forks.

....

....

.... **(2 marks)**

(ii) Sketch on the axes a graph to indicate how the compression of the spring varies from the moment the bike hits the bump until the oscillations have been completely damped after about 2 complete cycles of oscillation. **(4 marks)**

Compression / cm

Time / s

Had a go ☐ Nearly there ☐ Nailed it! ☐

Gravitational fields

1 A moon rock is brought to Earth. Which of its properties will change as a result?

☐ **A** mass ☐ **B** weight ☐ **C** density ☐ **D** volume **(1 mark)**

2 The Earth orbits the Sun because there is a gravitational force F from the Sun. What is the magnitude of the gravitational force acting on the Sun from the Earth?

☐ **A** 0 ☐ **B** less than F ☐ **C** F ☐ **D** greater than F **(1 mark)**

Guided

3 During the Apollo 14 mission, astronaut Alan Shepard hit a golf ball and claimed that it travelled for 'miles and miles'. Given that a good golfer on Earth can drive a ball over 250 m and that the gravitational field strength g at the surface of the Earth is $9.81\ \text{N kg}^{-1}$ and at the surface of the Moon is $1.63\ \text{N kg}^{-1}$, is it possible that Alan Shepard's claim was true? Your answer should be supported by relevant calculations. (1 mile is approximately 1.6 km).

Besides the reduced value of g, are there any other relevant differences on the Moon?

Assume that Alan Shepard hit the ball with the same initial speed as on Earth.

The horizontal component of the ball's velocity will not be affected but the reduced value of g affects the vertical component. Initially the vertical component of velocity will be the same as on Earth but it will take longer for the Moon's gravity to reduce this to zero and then bring it back down.

The time of flight (time to fall to the ground) increases by a factor of $\frac{9.81}{1.63}$ ....

....

....

....

.... **(4 marks)**

4 (a) State what is meant by a uniform gravitational field.

....

.... **(1 mark)**

(b) Under what circumstances can the gravitational field of the Earth be treated as uniform? (The radius of the Earth is 6400 km.)

....

.... **(1 mark)**

Newton's law of gravitation

1 The orbit of the planet Mars is elliptical and the planet's distance from the Sun varies. The diagram shows the shape of its orbit (the ellipse is exaggerated).

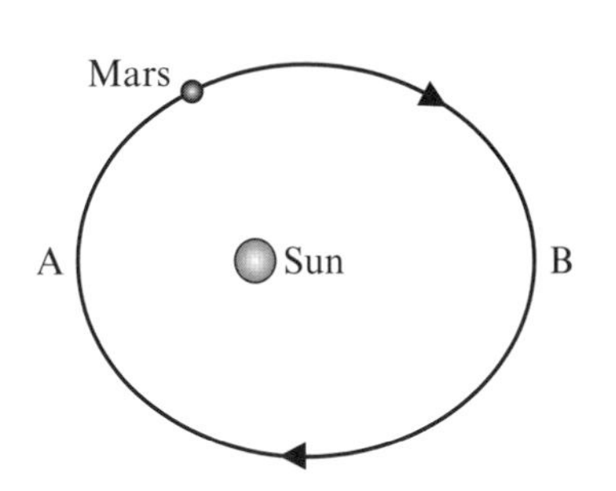

At their closest approach, the distance between the Sun and Mars is 207 million km.

At their farthest point, the distance is 249 million km.

(a) Calculate the ratio of the gravitational force on Mars at its point of closest approach to the Sun (position A) to the gravitational force on Mars at its farthest point from the Sun (position B).

....

.... **(2 marks)**

(b) Add an arrow to the diagram to show the direction of the resultant force on Mars at the position shown. **(1 mark)**

(c) Describe the changing speed and path of Mars at the position shown.

.... **(1 mark)**

(d) Describe how the gravitational potential energy and kinetic energy of Mars vary as it completes one orbit.

....

....

....

.... **(4 marks)**

(e) At which point in its orbit does Mars have its maximum speed?

.... **(1 mark)**

(f) Mars has a mass of 6.39×10^{23} kg and the Sun has a mass of 1.99×10^{30} kg. Calculate the maximum gravitational attraction between Mars and the Sun.

....

.... **(2 marks)**

Guided

2 The Moon has mass 7.35×10^{22} kg and radius 1740 km.

Calculate the gravitational field strength at the surface of the Moon.

Gravitational field strength $g = \frac{\text{gravitational force}}{\text{mass}} = \frac{F}{m}$

Gravitational force on a mass m is $F = \frac{GMm}{r^2}$ where M is the mass of the Moon and r is its radius.

....

.... **(2 marks)**

Had a go ☐ Nearly there ☐ Nailed it! ☐

Kepler's laws for planetary orbits

1 (a) State Kepler's second law.

....

.... **(1 mark)**

(b) Comets have very eccentric orbits (as shown in the diagram). Explain why they spend most of their time far from the Sun.

....

....

.... **(2 marks)**

2 Here is some incomplete data about the average radii and time periods of the inner planets in our solar system. Use Kepler's third law to complete the table.

Planet	Mean distance from the Sun / million km	Orbit time / years
Mercury	58	
Venus		0.62
Earth	150	1
Mars	228	

(3 marks)

Guided

3 The orbital period of the Earth's Moon is 27.3 days and its distance from the Earth is 380 000 km. Use this information to calculate the radius of orbit of a geostationary satellite (a satellite that orbits above a fixed point on the Earth's equator).

Kepler's third law will apply to all satellites of the same central body.

This means that r^3/T^2 is the same for the satellite as for the Moon.

For the satellite to be geostationary its period must be $T = 24$ hours.

....

....

.... **(3 marks)**

Satellite orbits

1 Which of the following relationships correctly links the orbital period, T, of an Earth satellite to its orbital radius, r?

☐ **A** $T \propto r^{-2}$ ☐ **B** $T \propto r^{1/2}$ ☐ **C** $T \propto r^{2/3}$ ☐ **D** $T \propto r^{3/2}$ **(1 mark)**

Guided

2 (a) Show that the time period for a satellite orbiting a planet of mass M at distance R does not depend on the mass m of the satellite.

Gravitational forces provide the centripetal force: $\frac{GMm}{r^2} = mr\omega^2$

What cancels from the equation?

And $\omega = \frac{2\pi}{T}$ ....

....

....

.... **(3 marks)**

(b) An Earth satellite is in a circular orbit with a period of 2.5 hours. Calculate the altitude (height above Earth's surface) of the satellite.
(Radius of Earth = 6400 km, mass of Earth = 6.0×10^{24} kg.)

....

....

....

.... **(3 marks)**

3 A spacecraft is in an elliptical orbit around the Earth similar to the one shown in the diagram.

spaceship

A

Earth

B

(a) Describe the energy transfers that take place as the spaceship completes one orbit starting and ending at point A.

....

....

.... **(3 marks)**

(b) Eventually the spacecraft is going to leave the Earth's orbit and travel to a distant planet. Discuss whether it will require more fuel to do this if it leaves the orbit at A or at B.

Consider the minimum energy needed to escape and the total energy of the spacecraft while in orbit.

....

....

....

.... **(4 marks)**

Had a go ☐ Nearly there ☐ Nailed it! ☐

Gravitational potential

1 (a) Explain what is meant by 'the gravitational potential at a point in space'.

....

.... **(2 marks)**

(b) Explain why the gravitational potential at any point in space is negative.

Think about the work that must be done to move a mass to infinity.

....

.... **(2 marks)**

Guided

(c) State the connection between gravitational field strength and gravitational potential.

.... is the (negative) gradient of the ....

.... The stronger the gravitational field strength, the more rapidly the potential changes with position. **(2 marks)**

(d) The graph shows how the magnitude of the gravitational field strength varies outside the Earth. Show, by drawing on the diagram, how the graph could be used to find the gravitational potential at a distance x from the centre of the Earth.

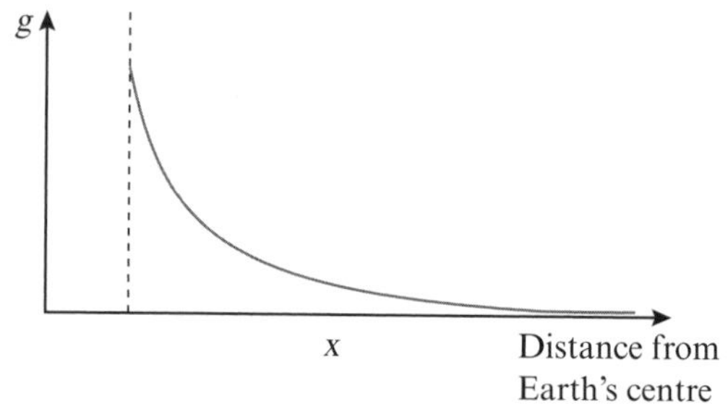

(3 marks)

(e) Explain how changes of gravitational potential energy of an object are related to gravitational potential.

....

.... **(2 marks)**

2 The diagram shows equipotentials close to the surface of the Earth. These mark areas in which an object would have the same gravitational potential energy. The vertical distance between adjacent equipotentials is 1.0 m.

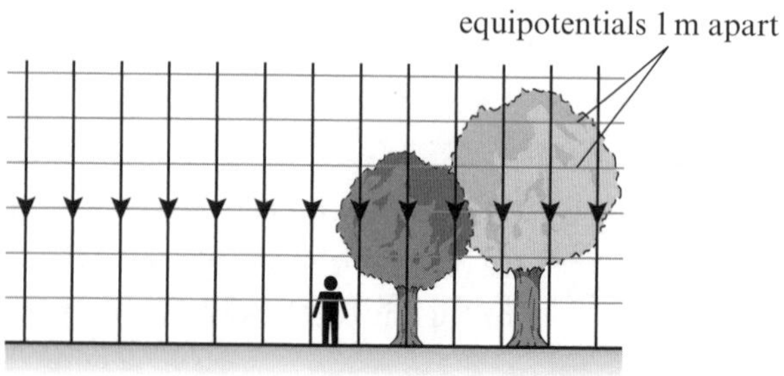

(a) Label the equipotentials with values of gravitational potential difference from the surface (taken to be zero).
The gravitational field strength near the surface is 9.81 $N\,kg^{-1}$. **(2 marks)**

(b) Explain why, in this example, it is reasonable to assume that the gravitational field strength is constant.

....

.... **(2 marks)**

Gravitational potential energy and escape velocity

1 The diagram shows some equipotentials close to the surface of the Earth.

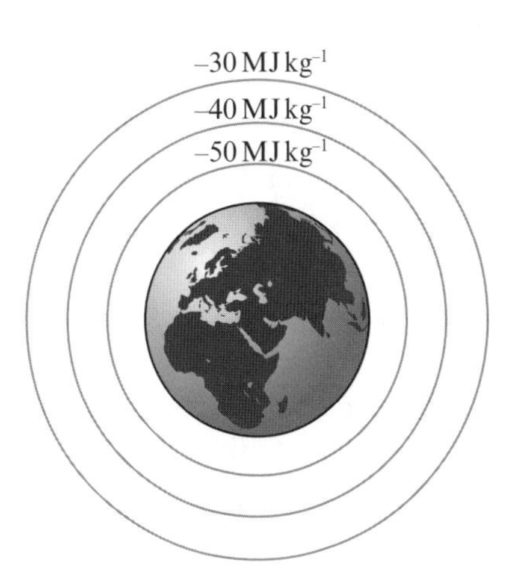

(a) A satellite of mass 1200 kg is to be moved from a circular orbit on the $-40\ \mathrm{MJ\,kg^{-1}}$ equipotential to a larger circular orbit on the $-30\ \mathrm{MJ\,kg^{-1}}$ equipotential. Calculate the change in its gravitational potential energy and state whether this is an increase or a decrease.

The work done is equal to the change in gravitational potential energy $W = m\Delta V_g$

$\Delta V_g = -30\ \mathrm{MJ\,kg^{-1}} - (-40\ \mathrm{MJ\,kg^{-1}}) =$ ....

....

.... **(3 marks)**

(b) The gravitational potential at the surface of the Earth is approximately $-60\ \mathrm{MJ\,kg^{-1}}$.

(i) Calculate the minimum kinetic energy required for a projectile of mass 500 kg, launched from the Earth's surface, to escape from the Earth's gravitational field.

....

....

.... **(2 marks)**

(ii) Calculate the escape velocity from the Earth's surface.

....

....

.... **(3 marks)**

(iii) Explain why it would not be possible to launch a spacecraft into deep space by giving it this initial velocity from the Earth's surface.

....

....

.... **(2 marks)**

Had a go ☐ Nearly there ☐ Nailed it! ☐

Exam skills

1 State Newton's law of gravitation.

....

.... **(1 mark)**

2 Define gravitational field strength and state its S.I. units.

....

.... **(2 marks)**

3 A satellite in a polar orbit has an orbital period of 2.0 hours. Calculate the altitude of the satellite above the surface of the Earth.

(Mass of Earth = 6.0×10^{24} kg; radius of Earth = 6400 km.)

....

....

....

.... **(4 marks)**

4 (a) What is meant by 'the gravitational potential at a point in space'?

....

.... **(2 marks)**

(b) Explain why all gravitational potential energies are negative.

....

....

.... **(3 marks)**

5 The Moon's orbital radius is increasing at a rate of about 38 mm per year. Its present value is 3.9×10^{8} m and the mass of the Moon is 7.3×10^{22} kg.

(a) Calculate the rate (in watts) at which the Moon's gravitational potential energy is increasing.

....

....

....

.... **(3 marks)**

(b) Suggest how the total energy of the Earth–Moon system can still be conserved despite the increase in the Moon's gravitational potential energy.

....

....

.... **(2 marks)**

Formation of stars

1 Which of the following energy transfers is the main one that takes place as a protostar forms from a collapsing gas cloud?

☐ **A** gravitational potential energy to nuclear potential energy

☐ **B** thermal energy to nuclear potential energy

☐ **C** gravitational potential energy to thermal energy

☐ **D** nuclear potential energy to gravitational potential energy. **(1 mark)**

2 Stars form when clouds of gas and dust collapse under their own gravitational fields.

(a) Our Sun has been in a relatively stable state for the past 5 billion years. Explain what stopped the collapse that formed it.

....

....

....

.... **(3 marks)**

(b) The Sun is currently 71% hydrogen and 27% helium (with small percentages of other light atoms). Explain how and why these percentages will change over the next 5 billion years.

....

....

....

.... **(3 marks)**

3 State **three** differences between a star and a planet.

....

....

.... **(3 marks)**

Guided **4** More massive stars have shorter lives. Suggest a reason for this.

Conditions at the core of a more massive star are ....
than in a less massive star. This greatly increases the rate of
.... reactions, reducing the lifetime of the star. **(2 marks)**

Had a go ☐ Nearly there ☐ Nailed it! ☐

Evolution of stars

1 Which of the following is **not** a stage in the life of a very massive star?

☐ **A** protostar ☐ **B** neutron star

☐ **C** white dwarf ☐ **D** red supergiant **(1 mark)**

Guided

2 (a) Explain what is meant by the 'Chandrasekhar limit'.

This is the maximum .... for a white dwarf star and is about .... times the mass of the Sun. More massive stars will collapse beyond the .... stage to form .... or even .... **(4 marks)**

(b) State what prevents a white dwarf star from collapsing to become a neutron star.

.... **(1 mark)**

(c) Suggest what might prevent a neutron star from collapsing to become a black hole.

.... **(1 mark)**

3 The diagram shows a NASA image of a planetary nebula.

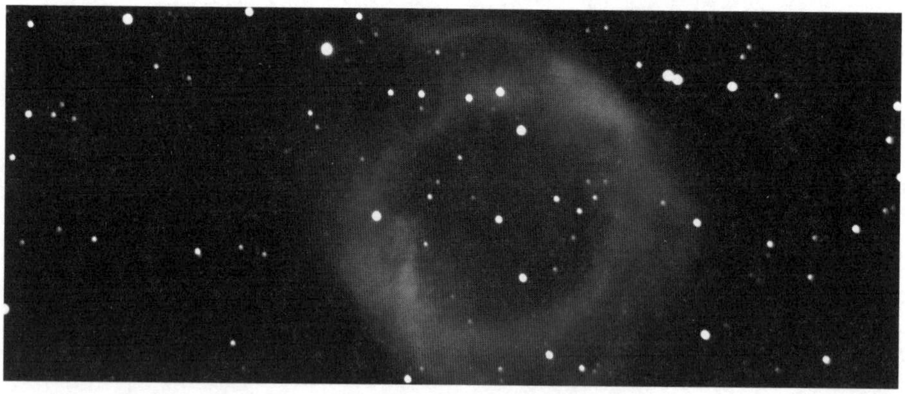

(a) Describe, by referring to the image, the main features of a planetary nebula.

....

....

.... **(2 marks)**

(b) Describe and explain the formation of a planetary nebula.

....

....

....

....

....

.... **(4 marks)**

End points of stars

1 Which of the following has the highest surface temperature?

☐ **A** a star like our Sun ☐ **C** a red giant star

☐ **B** a white dwarf star ☐ **D** a neutron star **(1 mark)**

2 The diagram below shows the current standard concept of all the stages in the Sun's life.

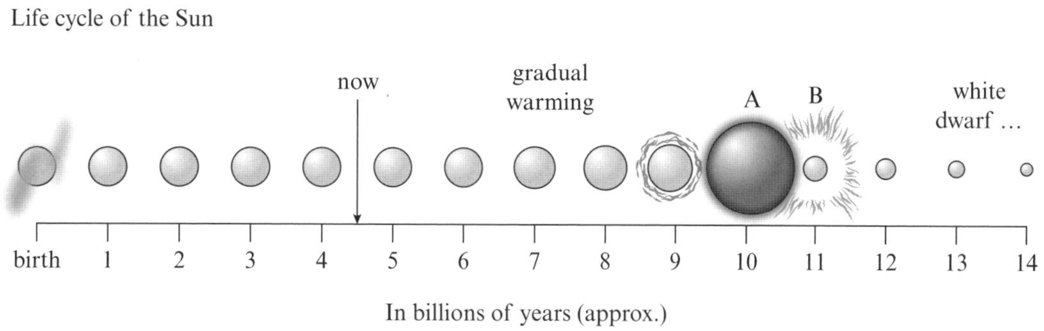

Life cycle of the Sun

now

gradual warming

A B

white dwarf …

birth 1 2 3 4 5 6 7 8 9 10 11 12 13 14

In billions of years (approx.)

(a) (i) Name the type of star labelled A.

.... **(1 mark)**

Guided (ii) Explain how the Sun changes into a star like the one at A.

Once most of the hydrogen in the core has ....

into helium, the radiation pressure .... The star

contracts under .... forces, and helium fusion

begins in the star's outer layers. The energy released in the helium

shell causes radiation pressure that makes the outer layer of the

star .... As the surface area of the star

increases, its surface temperature .... **(3 marks)**

(b) Name the object labelled B.

.... **(1 mark)**

(c) Explain why the Sun will become a white dwarf but not a neutron star.

....

.... **(3 marks)**

3 Under what circumstances will the core of a collapsing star become a black hole?

....

....

.... **(2 marks)**

Had a go ☐ Nearly there ☐ Nailed it! ☐

The Hertzsprung–Russell diagram

1 Explain what is meant by each of the following:

There are 2 marks per item, so try to make two points in each of your answers.

(a) stellar luminosity

....

....

(b) white dwarf star

....

....

(c) neutron star

....

....

(d) black hole

....

....

(e) main-sequence star

....

.... **(10 marks)**

Guided

2 The Hertzsprung–Russell diagram shows the life cycle of a star like our Sun.

Describe the changes that take place in stages 1 to 4.

Stage 1: gas cloud collapses to form ....

....

Stage 2: ....

Stage 3: ....

....

Stage 4: ....

.... **(6 marks)**

Energy levels in atoms

1 Which of the following electron 'jumps' in a hydrogen atom would result in the emission of the shortest wavelength radiation?

☐ **A** from $n = 5$ to $n = 3$

☐ **C** from $n = 2$ to $n = 7$

☐ **B** from $n = 4$ to $n = 1$

☐ **D** from $n = 1$ to $n = 2$

(1 mark)

2 The energy levels in the hydrogen atom are given by the equation $E = \frac{-13.6}{n^2}$ eV.

Guided

(a) Explain why all of the energy levels have negative values.

Electrons are bound to the .... so work has to be done to

.... from the nucleus. This raises their energy to

.... so the initial potential energy must have been negative.

(2 marks)

(b) Calculate the energy that must be supplied to move an electron from the ground state ($n = 1$) to the second excited state ($n = 3$). Give your answer in electronvolts.

....

....

....

(3 marks)

(c) Atoms absorb radiation strongly when the photons of that radiation have the same energy as an available quantum jump inside the atom. Calculate the wavelength of radiation that is absorbed when an electron is promoted from $n = 1$ to $n = 3$.

....

....

....

(3 marks)

(d) The ionisation energy for hydrogen is the minimum energy needed to eject an electron, initially in the ground state, from the atom. This is equivalent to raising the electron's potential energy to zero. Calculate the ionisation energy of an atom of hydrogen in electronvolts.

....

....

(2 marks)

(e) When two hydrogen atoms collide, the collision can be elastic or inelastic. Discuss, in terms of energy, the conditions under which the collision would be elastic.

....

....

....

(3 marks)

Emission and absorption spectra

1 White light with a continuous spectrum passes through a region of space containing a cool monatomic gas. What kind of spectrum will be observed in the transmitted light?

☐ **A** line emission spectrum ☐ **C** band emission spectrum

☐ **B** line absorption spectrum ☐ **D** band absorption spectrum **(1 mark)**

2 The diagram shows the spectrum of visible light from a nearby star (ignoring any red shift) and the hydrogen spectrum to the same scale.

star

hydrogen

(a) Explain how it is possible to tell that these are both absorption spectra.

....

....

.... **(2 marks)**

(b) Explain how an absorption spectrum is formed.

....

....

....

.... **(3 marks)**

3 A student carries out an experiment to analyse the spectrum of light from sodium vapour. His experimental set-up is shown in the diagram.

collimated beam of sodium light

1st-order diffraction pattern

diffraction grating

light detector moved along this line

The student measures the positions of maxima in the first-order spectrum and finds two maxima very close together. Here are some data from his experiment.

Diffraction grating: 400 lines per mm

Angular positions of maxima: 13.627° and 13.650°

Use this data to determine the main wavelengths of light emitted by the sodium light source.

....

....

....

....

.... **(4 marks)**

Wien's law and Stefan's law

1 The diagram shows the spectrum of cosmic background radiation measured by the COBE (Cosmic Background Explorer) satellite.

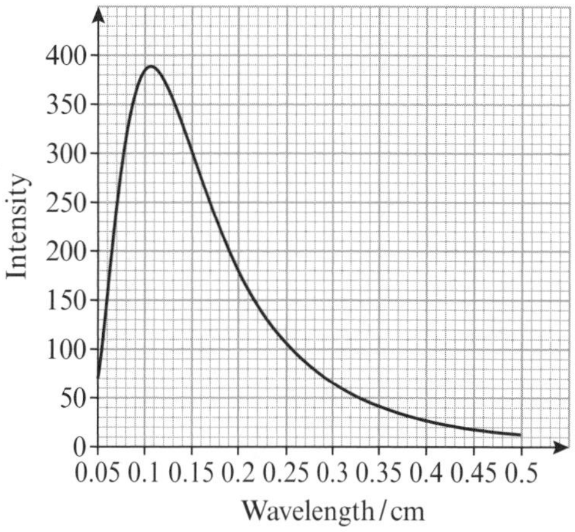

Cosmic background spectrum: COBE measurements.

(a) State the wavelength at which there is maximum intensity of background radiation.

....

(1 mark)

(b) The radiation is a close fit to a black-body radiation curve. Use your value from (a) to calculate the temperature of empty space.

You will need the constant in Wien's law, 2.9×10^{-3} m.K.

....

.... **(2 marks)**

(c) State which part of the electromagnetic spectrum this radiation belongs to.

.... **(1 mark)**

(d) Astronomers believe that the cosmic background radiation is a relic of the Big Bang. When the spectrum of this radiation is measured, it is found to be the same from all parts of the sky. What does this suggest about the thermal state of the early Universe? Give a reason for your answer.

....

.... **(2 marks)**

2 The star Betelgeuse has a surface temperature of 3500 K and a radius of 8.2×10^{11} m. The star Sirius has a surface temperature of 9940 K and a radius of 1.2×10^{9} m.

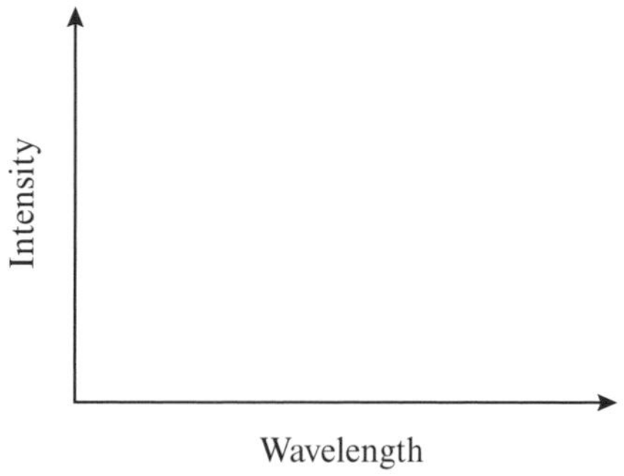

(a) Sketch, on the axes, the radiation spectra from Betelgeuse and Sirius. Include approximate values and units on the wavelength axis. Do **not** put values on the intensity axis. **(5 marks)**

Guided

(b) Calculate the total power radiated by Betelgeuse (its luminosity). Stefan's constant, $\sigma = 5.67 \times 10^{-8}\,\text{W m}^{-2}\,\text{K}^{-4}$.

Power radiated is given by Stefan's law: $L = 4\pi r^2 \sigma T^4$

....

.... **(2 marks)**

Had a go ☐ Nearly there ☐ Nailed it! ☐

The distances to stars

1 Long before Copernicus, the ancient Greek astronomer Aristarchus suggested that the planets, including the Earth, actually orbit the Sun. If this is the case, then the apparent positions of stars should shift as the Earth completes its orbit. However, these parallax effects were not observed until 1838.

Guided

(a) Suggest a possible reason why parallax effects were not observed until relatively recently.

If the distances to the nearest stars are very large then the parallax angles are small. It was therefore impossible for ancient astronomers to ....

.... **(2 marks)**

(b) How does the size of these parallax movements depend on the distance from the Earth to the star?

....

.... **(2 marks)**

(c) The first star for which the parallax angle was measured was 61 Cygni, by Bessel in 1838. His result was a parallax half angle of 0.314 arcseconds (1.52×10^{-6} radians). Use this value to calculate the distance from Earth to 61 Cygni, giving your answer in parsecs, metres and light years.

....

....

....

.... **(4 marks)**

(d) Earth-based telescopes can measure parallax angles for stars out to a distance of about 100 light years. However, the Hipparcus satellite can measure the parallax angles of stars out to about 1000 light years.

(i) Calculate the minimum size of parallax angles that can be measured from the Earth and from the Hipparcus satellite.

....

.... **(2 marks)**

(ii) Approximately how many times more stellar distances can be measured by Hipparcus than by ground-based telescopes? State any assumption that you make to answer this question.

Maths skills: The volume of a sphere is $\frac{4}{3}\pi r^3$

....

....

....

.... **(3 marks)**

The Doppler effect

1 A Formula 1 car races past a spectator at constant speed v as shown in the diagram. The spectator notices that the engine note (the frequency of the sound) seems to change as the car passes.

A B C
● spectator

(a) Explain why the frequency of the engine note heard by the spectator changes.

....

....

.... **(2 marks)**

Guided

(b) Explain why the engine note heard by the driver of the car does not change as the car moves from A to C.

The Doppler effect depends on the relative velocity of the source of sound to the person hearing the sound. Since the engine and driver are ....

.... **(2 marks)**

(c) (i) On the axes, sketch how the engine note heard by the spectator changes as the car moves from A to C. The dotted line represents the note heard by the driver.

Frequency of engine note heard by spectator

A B C

Car position

Think of the car's velocity as a vector. Does it have a component toward or away from the spectator?

Explain each section: A to B to B, B to C. **(3 marks)**

(ii) Explain the shape of your graph.

....

....

....

.... **(3 marks)**

2 Radio telescopes can detect radio waves from stars in the spiral arms of our galaxy. These have a characteristic wavelength of 21.106 cm if they are received from a stationary source. In one particular observation, the waves received on Earth are found to have a wavelength of 21.132 cm.

What can you conclude about the motion of these stars? Include a calculation.

....

....

.... **(2 marks)**

Had a go ☐ Nearly there ☐ Nailed it! ☐

Hubble's law

1 Which of the following statements about the theory of the expanding Universe is correct?

☐ **A** All galaxies are moving away from us, so the Earth is in a special position at the centre of the Universe.

☐ **B** Hubble's law shows that very distant galaxies have the same recession velocities as nearby galaxies.

☐ **C** Both red shifts and recession velocities increase with distance.

☐ **D** Light from distant galaxies is red shifted but light reaching those galaxies from the Earth would be blue shifted. **(1 mark)**

Guided

2 (a) Explain how the red shift of light from a distant galaxy can be used to estimate the distance to that galaxy. State any other pieces of information you would need in order to carry out this estimate.

Red shift $z = \frac{\Delta\lambda}{\lambda_0}$ is related to recession velocity v by the equation $z = \frac{v}{c}$. Recession velocity is related to distance by Hubble's law: $v = H_0 d$, so ....

....

....

....

.... **(4 marks)**

(b) Suggest two reasons why the result of your calculation in (a) would be an estimate rather than an accurate value.

....

....

.... **(2 marks)**

3 (a) State Hubble's law as an equation and define the terms in it.

....

....

....

.... **(3 marks)**

(b) The current best value for the age of the Universe is 13.7 billion years. Use this value to estimate the Hubble constant, including an appropriate unit.

....

....

.... **(2 marks)**

The evolution of the Universe

1 Which statement about the evolution of the Universe is correct?

☐ **A** An open Universe will collapse again in a 'Big Crunch' at some time in the future.

☐ **B** A closed Universe will expand forever.

☐ **C** Dark energy will eventually stop the expansion of the Universe.

☐ **D** The greater the current rate of expansion, the less time since the Big Bang. **(1 mark)**

Guided

2 (a) Explain how detailed observations of the rotation of galaxies led cosmologists to propose the existence of 'dark matter'.

All galaxies rotate. The .... that keeps stars in orbit around the galactic nucleus is provided by ....

However, the total .... attraction from all the visible matter is not enough to provide the required force so dark matter

.... **(4 marks)**

(b) State **two** characteristics of dark matter.

....

.... **(2 marks)**

3 (a) State Hubble's law.

.... **(1 mark)**

(b) The Hubble constant is $H_0 = 71\,\text{km}\,\text{s}^{-1}\,\text{Mpc}^{-1}$ ($1\,\text{pc} \approx 3.1 \times 10^{16}\,\text{m}$). Express the Hubble constant in S.I. units.

....

.... **(2 marks)**

Maths skills

(c) Calculate the recession velocity of a galaxy 100 MPc from our own Milky Way galaxy.

....

.... **(2 marks)**

(d) Use your answer to (c) to estimate the time in years since the distant galaxy and the Milky Way were at the same point in space and comment on this.

....

....

....

.... **(2 marks)**

Capacitors

1 Which of the following combinations of units is **not** equivalent to the farad?

☐ **A** $\mathrm{A\,s\,V^{-1}}$ ☐ **B** $\mathrm{C\,V^{-1}}$ ☐ **C** $\mathrm{C^2\,J}$ ☐ **D** $\mathrm{s\,\Omega^{-1}}$ **(1 mark)**

2 The graph shows how the charge stored on a capacitor increases as the p.d. across it is increased.

(a) Calculate the capacitance of the capacitor.

....

(2 marks)

(b) (i) What aspect of the graph represents the energy stored on the capacitor?

.... **(1 mark)**

Guided

(ii) Calculate the energy stored when the capacitor is charged to 40 V.

area of a triangle $= \frac{1}{2} \times$ base $\times$ height

energy $= \frac{1}{2} \times$ .... $\times$ ....

.... **(2 marks)**

(iii) Calculate the energy stored when the capacitor is charged to 20 V.

....

.... **(2 marks)**

3 A 220 µF capacitor is charged from a 12 V supply through a 100 Ω resistor using the circuit shown below.

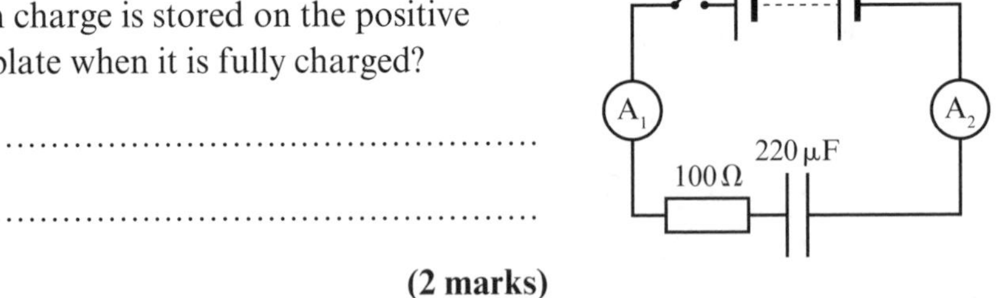

(a) How much charge is stored on the positive capacitor plate when it is fully charged?

....

....

(2 marks)

(b) What current flows through each ammeter at the moment just after S is closed?

....

.... **(2 marks)**

(c) Explain why the current will decrease as the capacitor charges up.

....

....

.... **(2 marks)**

Series and parallel capacitor combinations

Guided

1 Two identical capacitors of capacitance C can be connected in series or in parallel. Their total capacitance when connected in parallel is C_p and when connected in series is C_s. What is the value of the ratio $\frac{C_p}{C_s}$?

> Write out the equations for capacitance of two capacitors in series and in parallel, then the ratio. You can see immediately that C is incorrect.

☐ **A** $\frac{1}{4}$ ☐ **B** $\frac{1}{2}$ ☑ **C** 2 ☐ **D** 4 **(1 mark)**

2 Two capacitors are connected into the circuit shown. Both switches are initially open.

6.0 V
S_1 22 μF
S_2 47 μF

(a) Switch S_1 is closed but S_2 remains open. Calculate the charge stored on the 22 μF capacitor.

....

.... **(2 marks)**

(b) Switch S_1 is now opened and S_2 is closed. Calculate the charge stored on the 47 μF capacitor.

....

.... **(2 marks)**

(c) Explain why no further charge will flow if S_1 is now closed (so that both switches are now closed).

> Current will only flow between two points if there is a potential difference.

....

.... **(2 marks)**

(d) State the total charge stored on the two capacitors and the potential difference across each of them.

....

.... **(2 marks)**

(e) Calculate the total capacitance of a 22 μF capacitor connected in parallel with a 47 μF capacitor and show that it is the same as the ratio of charge to potential difference in part (d).

....

....

.... **(2 marks)**

(f) The same two capacitors are connected in series to the 6.0 V power supply.

Calculate their total capacitance.

> When two capacitors are in series they each store the same charge.

....

.... **(2 marks)**

Had a go ☐ Nearly there ☐ Nailed it! ☐

Capacitor circuits

1 (a) Explain the significance of the time constant $\tau = RC$ in a capacitor-charging circuit.

....

....

.... **(3 marks)**

(b) Show that the unit of resistance × the unit of capacitance is the unit of time.

> **Maths skills** Units in all physical equations must balance. Use equations you know to reduce complex unit such as the ohm to simpler combinations (e.g. $1\ \Omega = 1\ \mathrm{V\,A^{-1}}$).

....

....

.... **(2 marks)**

2 A capacitor is charged using the circuit shown in the diagram.

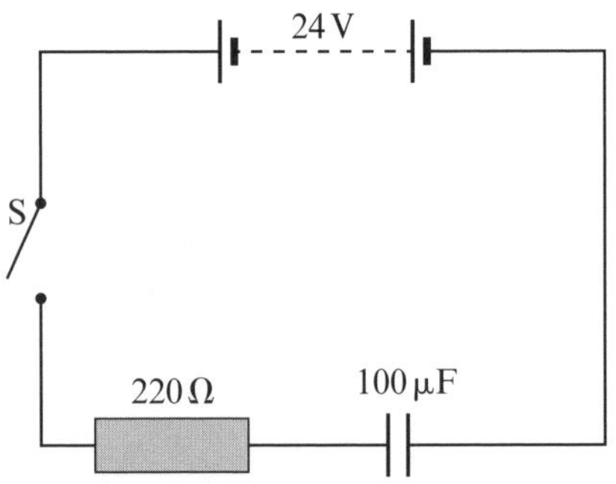

24 V

S

220 Ω

100 μF

(a) Calculate the time constant for this circuit.

....

....

(2 marks)

(b) S is closed and the capacitor begins to charge up. Approximately how long will it take for the capacitor to reach 99% of its full charge?

> A useful rule of thumb is that over one time constant the charge on a charging/discharging capacitor changes by 37%.

....

....

.... **(2 marks)**

Guided

(c) The time constant and therefore the time taken to charge the capacitor increases if the capacitance or resistance is increased. Explain why this is.

Increasing C increases the final charge ($Q = CV$) so ....

....

Increasing R decreases the charging current so ....

.... **(4 marks)**

3 A charged capacitor of capacitance 50 μF is discharged through a resistor of resistance 20 kΩ. Calculate the time taken for the charge on the capacitor to fall to approximately 1% of its initial value.

> First convert the units μF and kΩ to F and Ω.

....

....

.... **(2 marks)**

Exponential processes

1 Which of the following statements about the discharge of a capacitor through a fixed resistance is **not** correct?

☐ **A** The time taken for the voltage to fall to half of any initial value is always the same.

☐ **B** The larger the resistance or capacitance, the longer it takes to discharge.

☐ **C** The charge, current and voltage all have the same time constant for decay.

☐ **D** The rate at which the charge leaves the capacitor is constant. **(1 mark)**

2 The 1000 μF capacitor shown in the diagram is charged for 3.0 s and then discharged through a 500 Ω resistor.

6.0 V

1000 μF

500 Ω

(a) Use the axes to sketch a graph to show how the potential difference across the capacitor changes with time during the charging and discharging cycle. **(6 marks)**

6.0

p.d / V

0 3.0 6.0

Time / s

(b) Calculate the charge on the capacitor when it is fully charged.

....

(1 mark)

Maths skills Calculate the time constant and use this as a 'yardstick' for the time scale. How long does the capacitor take to charge/discharge?

Guided

(c) Calculate the maximum charging current and state when it occurs.

Charging current I is equal to $\frac{V_R}{R}$, where V_R is the potential difference across the resistor.

$V_R = 6.0 - V_C$ so will be maximum when the capacitor is uncharged ($V_C = 0$). ....

.... **(2 marks)**

(d) How long will it take for the charge on the discharging capacitor to fall from its initial value to half that value?

Maths skills Take the natural logarithm of both sides of the equation to solve it for t.

....

....

.... **(3 marks)**

(e) Add a line to your graph to show how the potential difference across the resistor changes during the charging and discharging cycle. **(2 marks)**

Exam skills

1 The circuit shown in the diagram is a simplified model of the circuit used in a camera flash gun. The lamp emits light when the potential difference across it is greater than 60 V.

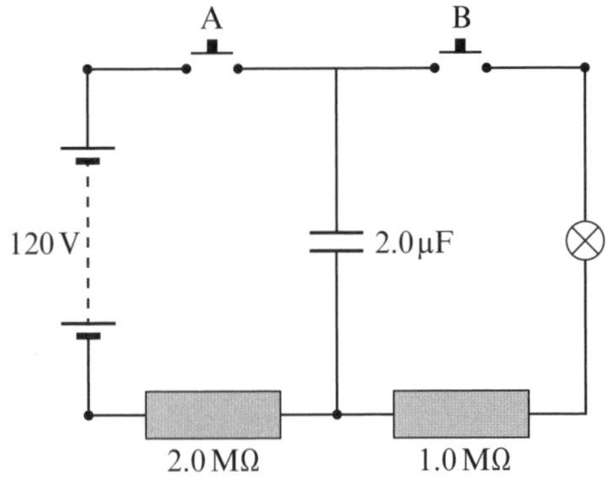

The sequence of events to make the lamp flash is:

(1) close switch A for at least 12 s
(2) open switch A
(3) close switch B
(4) open switch B.

A
B
120 V
2.0 μF
2.0 MΩ
1.0 MΩ

(a) Explain why it is recommended to close switch A for at least 12 s. Support your answer with a relevant calculation.

....

....

....

.... **(3 marks)**

(b) Calculate the energy stored on the capacitor when:

(i) it is fully charged by the 120 V supply

....

.... **(2 marks)**

(ii) its voltage has fallen to 60 V.

....

.... **(1 mark)**

(c) Calculate the time taken for the voltage across the capacitor to fall from 120 V to 60 V during its discharge. Assume that the resistance of the lamp can be neglected.

....

.... **(2 marks)**

(d) Use your answers to (b) and (c) to estimate the power output of the flash lamp.

....

....

....

.... **(4 marks)**

(e) The owner of the camera notices that when he uses the flash twice in succession it takes less time to charge up after the first use. Suggest a reason for this.

....

....

.... **(2 marks)**

Electric fields

1 Two charged spheres are placed a distance d apart. Which of the following changes could leave the force between the spheres unchanged?

☐ **A** Moving the charges further apart and increasing the magnitude of charge on one of the spheres.

☐ **B** Changing the sign of charge on both spheres and moving them closer together.

☐ **C** No change to either charge but moving the charges farther apart.

☐ **D** Changing the sign of charge on one sphere but keeping them at the same distance from one another. **(1 mark)**

2 Many molecules are polar. This means that one end of the molecule is slightly positive and the other end is slightly negative. Sketch the electric field around the polar molecule in the diagram below.

+ –

(3 marks)

3 Explain why, when an electric field is represented by field lines, the field lines cannot cross one another.

....

....

.... **(2 marks)**

4 When Geiger and Marsden investigated the structure of the atom, they used alpha particles as projectiles and realised that they were being deflected by the electrostatic force from a dense region of charge, the gold nucleus. An alpha particle has charge $+4e$ and a gold nucleus has a charge $+79e$. Which statement gives the correct relationship between the magnitudes of the forces on the alpha particle and gold nucleus when they interact?

☐ **A** They are equal.

☐ **B** The force on the alpha particle is $\frac{79}{4}$ times greater than the force on the gold nucleus.

☐ **C** The force on the gold nucleus is $\frac{79}{4}$ times greater than the force on the alpha particle.

☐ **D** It is impossible to say because it depends on the distance between them. **(1 mark)**

Had a go ☐ Nearly there ☐ Nailed it! ☐

Coulomb's law

1 Two particles, each carrying a charge Q, are separated by a distance d. The electrostatic force between them is F. What is the electrostatic force between two different particles that each carry a charge $2Q$ and are separated by a distance $2d$?

☐ **A** $\frac{F}{4}$ ☐ **B** $\frac{F}{2}$ ☐ **C** F ☐ **D** $2F$ **(1 mark)**

2 Many molecules are dipoles, positive at one end and negative at the other. A simple model of a dipole consists of two point charges $+Q$ and $-Q$ separated by a distance $2r$ as shown in the diagram.

X Y Z
$+Q$ $-Q$
r $2r$ r

(a) Calculate the force on an electron at Y (halfway between the two charges). Assume that $Q = 8.0 \times 10^{-20}$ C. (The charge on an electron is 1.6×10^{-19} C; $r = 2.5 \times 10^{-10}$ m; $\varepsilon_0 = 8.85 \times 10^{-12}$ F m^{-1})

....

....

....

.... **(4 marks)**

(b) Calculate the force on an electron at Z.

The electron will be affected by both +Q and –Q.

....

....

....

.... **(4 marks)**

Guided

(c) Some solid materials contain dipole molecules. Suggest how these molecules might behave when an external electric field is applied to the solid. You might find it helpful to draw a diagram.

The external field will exert a force on each end of the dipole.

The charges at each end of the dipole are opposite so the forces will be in opposite directions.

Add arrows to the diagram to show the directions of the forces on each charge.

+
–

Unless the molecule is already aligned with the field, these two forces will create a turning effect. As a result ....

....

....

.... **(4 marks)**

Similarities between electric and gravitational fields

1 Which of the following statements about electric and gravitational fields is incorrect?

☐ **A** All masses accelerate at the same rate in a uniform gravitational field.

☐ **B** All charges accelerate at the same rate in a uniform electric field.

☐ **C** The electric field strength at a point is the vector sum of all electric fields superposed at that point.

☐ **D** The gravitational field strength at a point is the vector sum of all gravitational fields superposed at that point. **(1 mark)**

2 It is now known that in Rutherford's scattering experiment, positively charged alpha particles scattered from positively charged gold nuclei because of the electrostatic repulsions between the two particles. However, there was also a gravitational attraction between the particles.

If you approach this task algebraically you will be able to cancel out the particle separation.

Use the data below to show that Rutherford was justified in neglecting gravitational effects.

Mass of alpha particle: 6.64×10^{-27} kg   Charge on alpha particle: 3.20×10^{-19} C

Mass of gold nucleus: 3.27×10^{-25} kg   Charge on gold nucleus: 1.27×10^{-17} C

....

....

....

....

.... **(4 marks)**

3 (a) Draw one arrangement of charges that has a positive electrostatic potential energy and explain why the electrostatic potential energy is positive.

....

....

.... **(3 marks)**

(b) Explain why it is not possible to create a positive gravitational potential energy.

....

....

.... **(2 marks)**

Uniform electric fields

1 Which of the following changes would **increase** the electric field strength between two overlapping parallel metal plates connected to a d.c. power supply?

☐ **A** increasing the separation between the plates

☐ **B** using plates with a greater area of overlap

☐ **C** decreasing the separation of the plates

☐ **D** using plates with smaller area of overlap. **(1 mark)**

Guided

2 (a) The air breaks down and becomes conducting when the electric field exceeds about $3 \times 10^6\,V\,m^{-1}$. In an experiment to demonstrate electrostatic effects, a teacher gradually increases the potential difference between two metal conductors until a spark jumps between them. The separation of the conductors is 2.5 mm. What is the approximate value of the minimum potential difference needed to produce the spark?

The field between the conductors might not be uniform so the equation $E = \frac{V}{d}$ is only approximately correct.

The air will break down when $\frac{V}{d} > 3 \times 10^6\,V\,m^{-1}$

....

.... **(2 marks)**

(b) A capacitor consists of two parallel metal plates separated by a distance of 4.0 mm. A fixed potential difference of 500 V is applied to the plates and they are gradually moved further apart until their separation is 8.0 mm.

(i) Calculate the electric field strength between the plates when they are 4.0 mm apart.

....

.... **(1 mark)**

(ii) Calculate the electric field strength between the plates when they are 8.0 mm apart.

....

.... **(1 mark)**

(iii) Sketch a graph to show how the electric field strength between the plates varies with their separation while the potential difference between them remains at 500 V, and label it with the values calculated above.

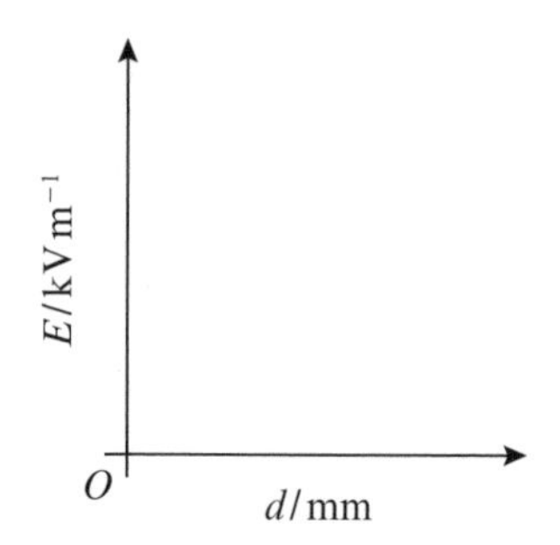

(3 marks)

Charged particles in uniform electric fields

1 A beam of electrons in a vacuum passes through a region of vertical uniform electric field of strength E between two charged parallel metal plates as shown in the diagram. The initial speed of the electrons is v and the length of the plates is d. The charge on an electron is e and the mass of an electron is m.

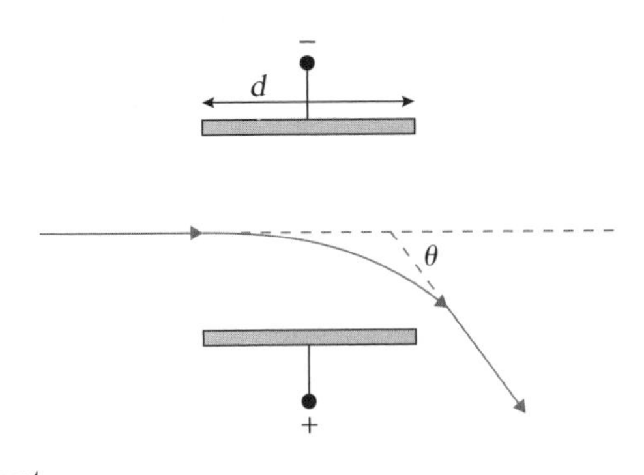

(a) Which of the following statements about the velocity of the electrons is true?

☐ **A** The vertical component of velocity is constant.

☐ **B** The horizontal component of velocity is constant.

☐ **C** There is a constant centripetal force on the electrons when they are between the plates.

☐ **D** The electric field does no work on the electrons. **(1 mark)**

(b) Write down an expression for the time t taken for an electron to pass between the plates (i.e. to move a distance d while it is between the plates).

.... **(1 mark)**

(c) Write down an expression for the acceleration of an electron when it is between the plates and state the direction of this acceleration.

....

.... **(2 marks)**

(d) Use your answers to (b) and (c) to derive an expression for the vertical displacement of the beam as it passes between the plates.

Use the SUVAT equations.

....

.... **(2 marks)**

(e) Derive an expression for the vertical component w of the electron's velocity as it leaves the region between the plates.

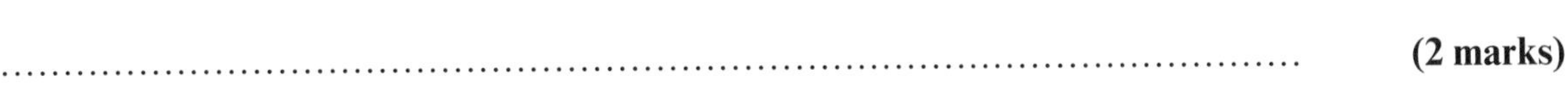

....

.... **(2 marks)**

(f) Derive an expression for the angular deflection, θ, of the beam as it passes between the plates.

We need the direction of the velocity vector as the electron leaves the field.

The horizontal component is v. The vertical component is $w =$ ....

$\tan\theta = \frac{w}{v} =$ ....

$\theta =$ ....

.... **(3 marks)**

Electric potential and electric potential energy

1 Ion X has mass m and charge q. Ion Y has mass $2m$ and charge q. Both ions are accelerated through the same potential difference V. Which line in the table gives the correct ratios for their kinetic energies (KE) and velocities (v)?

	KE_X/KE_Y	v_X/v_Y
☐ **A**	1	$\sqrt{2}$
☐ **B**	1	2
☐ **C**	$\frac{1}{2}$	$\sqrt{2}$
☐ **D**	$\frac{1}{2}$	2

(1 mark)

2 The spherical dome of a Van der Graaff generator is charged to a potential of 160 kV with respect to earth. The radius of the dome is 10 cm.

(a) Write down an expression for the electrical potential at a distance r from a point charge of magnitude Q.

.... **(1 mark)**

(b) The field and potential outside a charged sphere are exactly the same as for a point charge of the same magnitude placed at the centre of the sphere. Use this information to calculate the potential at:

(i) a point in space 10 cm above the surface of the dome.

.... **(1 mark)**

(ii) a point in space 30 cm above the surface of the dome.

.... **(1 mark)**

Guided

(c) Sketch on the axes, a graph to show how the electric potential varies from the surface of the dome to a point 30 cm above the surface of the dome.

(4 marks)

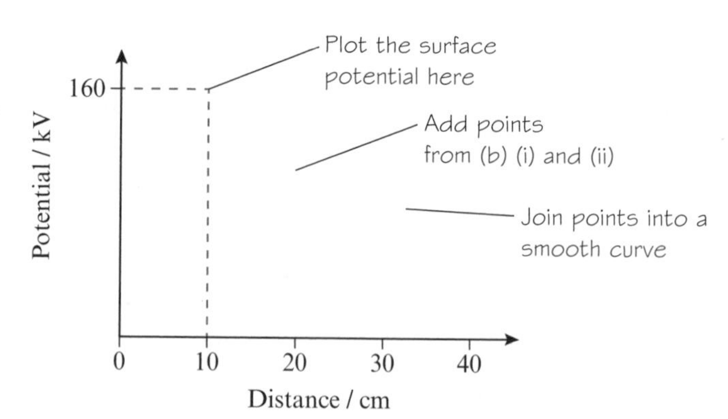

3 When a uranium-238 nucleus undergoes nuclear fission it separates into two daughter nuclei that are both positively charged. These fly apart because of their electrostatic repulsion. Estimate the energy released as the two daughter nuclei separate if they both have charges of about $46e$ and separate from an initial distance of about 5×10^{-15} m. Express your answer in joules and in electronvolts.

....

....

....

.... **(3 marks)**

Capacitance of an isolated sphere

1 The spherical dome of a Van der Graaff generator is charged up to a potential of 75 kV with respect to earth. Which line in the table gives the correct values for the electrical potential at the centre and at the surface of the dome?

	Centre	Surface
☐ **A**	0	0
☐ **B**	0	75 kV
☐ **C**	75 kV	0
☐ **D**	75 kV	75 kV

(1 mark)

Maths skills

2 An isolated, charged, conducting sphere of radius 2.0 cm is connected to the positive terminal of a 5000 V power supply.

(a) Calculate the charge stored on the sphere.

....

.... **(2 marks)**

(b) (i) Calculate the potential difference between two points A and B at distances 10 cm and 15 cm, respectively, from the centre of the sphere.

....

....

Use the equations for a radial field: $V \propto 1/r$

.... **(4 marks)**

Guided

(ii) A second conducting sphere has a charge of +2.0 nC and is placed at B. Calculate the work needed to move it from B to A.

Work done in an electric field is given by $Q\Delta V = Q(V_A - V_B)$

....

.... **(4 marks)**

Maths skills

3 A conducting sphere of radius 5.0 cm is connected momentarily to a 2000 V power supply. It is then connected to earth through a 22 kΩ resistor.

(a) Calculate the capacitance of the sphere.

.... **(2 marks)**

(b) Calculate the energy stored on the capacitor.

.... **(2 marks)**

(c) Calculate the initial discharge current.

....

.... **(2 marks)**

(d) Discuss whether or not the capacitor is fully discharged after 50 ms.

....

....

.... **(4 marks)**

Had a go ☐ Nearly there ☐ Nailed it! ☐

Representing magnetic fields

1 Which of the following statements about magnetic fields is correct?

☐ **A** The two ends of a bar magnet carry opposite charges.

☐ **B** Like poles attract one another.

☐ **C** Magnetic field lines go from south to north.

☐ **D** The stronger the magnetic field the denser the field lines. **(1 mark)**

Guided **2** (a) Explain what is meant by hard and soft magnetic materials.

Soft iron is a soft magnetic material.

A hard magnetic material is difficult to ....

or .... A soft magnetic material is easy to

.... and .... **(2 marks)**

(b) Strong electromagnets are used in car breakers' yards to lift vehicles and then drop them in a crushing machine. Suggest, with reasons, why a soft magnetic material such as iron is used for the core of such an electromagnet.

....

....

....

.... **(4 marks)**

3 The diagram below shows a wire carrying electric current into the page placed between the north and south poles of two permanent magnets. Add field lines to the diagram to show the resultant magnetic field between the poles and the direction of the force on the wire..

N ⊗ S

(4 marks)

4 A magnetic compass consists of a magnetised needle pivoted at its centre. Use a diagram to help to explain why it will always line up with the Earth's magnetic field.

Draw a uniform field from the Earth and show how this exerts forces on the poles of the magnetised needle. When will these forces be in equilibrium?

(4 marks)

Force on a current-carrying conductor

1 When current flows around a rectangular coil there is a repulsive force between opposite sides of the coil as shown in the diagram. Which of the following statements explains these forces?

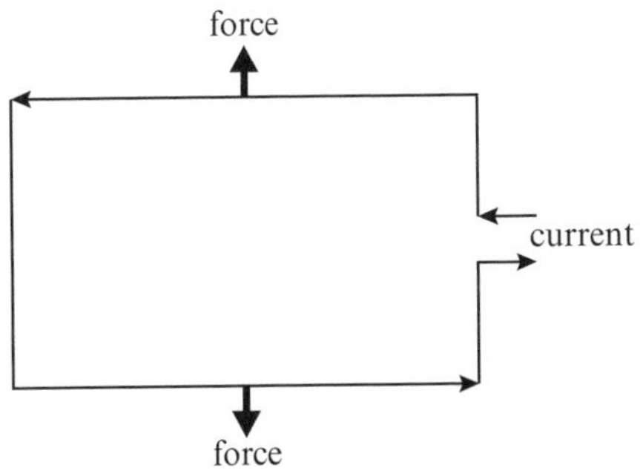

☐ **A** Like charges repel.

☐ **B** Opposite sides of the coil create opposite magnetic poles.

☐ **C** Currents create magnetic fields that interact with other currents.

☐ **D** The current in each wire is the same because it is a series circuit. **(1 mark)**

2 A student carries out an experiment to measure the force exerted on a wire carrying current through a magnetic field. The apparatus is shown in the diagram.

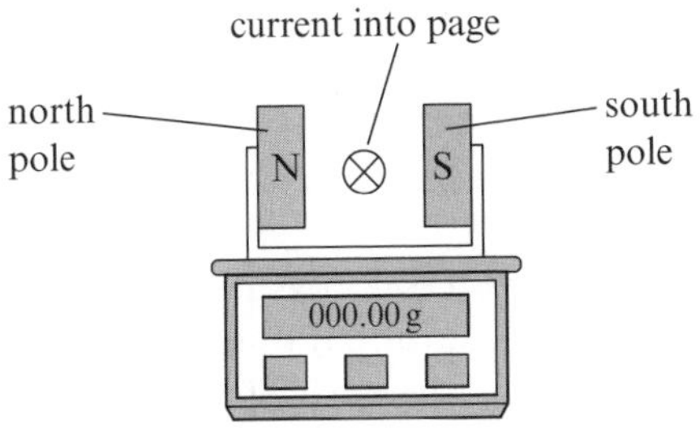

(a) Draw an arrow onto the diagram to show the direction of the magnetic force on the current-carrying conductor. **(1 mark)**

Guided

(b) State whether the balance reading will increase or decrease when a current is switched on and travels into the page.

The force on the magnet will be equal and opposite to the force on the wire (by Newton's third law).

The force on the magnet therefore acts in an .... direction.

This force will .... the contact force of the magnet on the balance so the reading on the balance will .... **(3 marks)**

(c) A current of 2.0 A is passed through the wire and the balance reading changes by 0.61 g.

(i) Calculate the magnitude of the magnetic force on the wire.

....

.... **(2 marks)**

(ii) The length of wire perpendicular to the magnetic field is 0.060 m. Calculate the magnitude of the magnetic flux density between the two poles (assume the flux density is uniform).

....

.... **(2 marks)**

Charged particles in a region of electric and magnetic field

1 An electron beam passes through a region of space in which there is both a uniform electric field E and a uniform magnetic field B. The directions of the two fields are perpendicular to one another and to the direction of the electron beam. The velocity of the electrons is v. Which condition below would result in no deflection for the electron beam?

☐ **A** $E = Bev$ ☐ **B** $E = vB$ ☐ **C** $E = \frac{B}{v}$ ☐ **D** $E = \frac{B}{e}$ **(1 mark)**

2 An electron moving at velocity v is fired into a uniform magnetic field acting into the page as shown in the diagram.

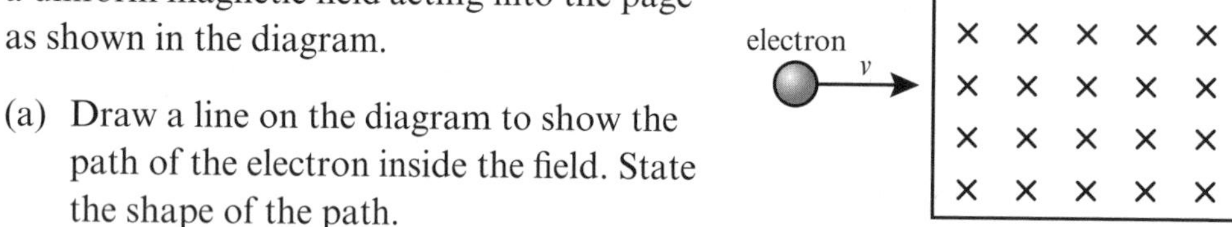

(a) Draw a line on the diagram to show the path of the electron inside the field. State the shape of the path.

.... **(2 marks)**

(b) Add a second line to show the path of a second electron that enters the field along the same line but with velocity $2v$. Explain why this path differs from the path of the first electron.

....

.... **(2 marks)**

Guided

3 A magnetic field can be used to separate isotopes, e.g. to separate fissile uranium-235 from non-fissile uranium-238. The first stage is to accelerate ions of the mixture up to the same speed. These ion beams are then injected into a uniform magnetic field at right angles to their motion. Inside the field they move in circular paths of different radii. In a particular experiment, the ions have velocity v and the field has strength B. The masses of the two ions are 235 u and 238 u (where u is the unified atomic mass unit) and they each carry a charge q.

Derive an expression for the separation of the ion beams after they have turned through a semicircle in the magnetic field.

For a charge Q and mass M moving at speed v perpendicular to a magnetic field of strength B, $\frac{Mv^2}{r} = BQv$, so $r = \frac{Mv}{BQ}$

For U-235, $r_{235} = \frac{235vu}{Bq}$

For U-238, $r_{238} =$ ....

The separation after a semicircle will be $2r_{238} - 2r_{235} =$ ....

....

.... **(4 marks)**

Magnetic flux and magnetic flux linkage

1 A uniform coil of N turns and area A lies with its plane parallel to a uniform magnetic field of strength B. Which of A–D gives the flux φ through the coil?

☐ **A** $\varphi = 0$ ☐ **B** $\varphi = B$ ☐ **C** $\varphi = BA$ ☐ **D** $\varphi = NBA$ **(1 mark)**

Guided

2 The Earth's magnetic field strength at a particular position near the surface has a strength of 36 μT at an angle of 30° to the horizontal.

Calculate the magnetic flux through 2.0 m² of the Earth's surface.

The vertical component of the magnetic field strength is

$B_{vert} = 36 \times 10^{-6} \times \sin 30°$.

The flux is given by $\varphi = B_{vert}A =$ ....

.... **(3 marks)**

3 The diagram shows a flat coil of 50 turns and area 80 cm² in a uniform magnetic field of flux density 0.22 T.

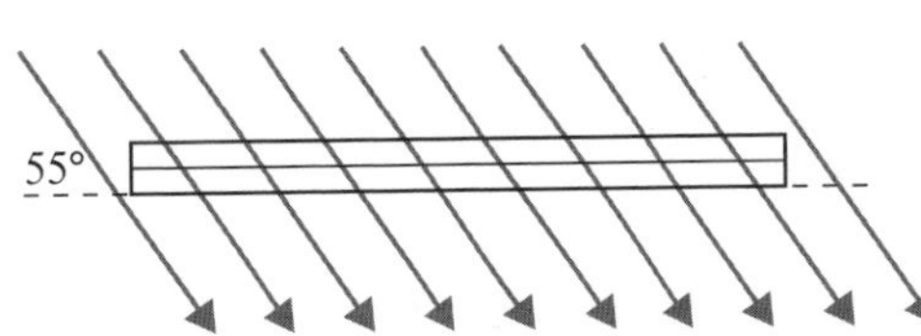

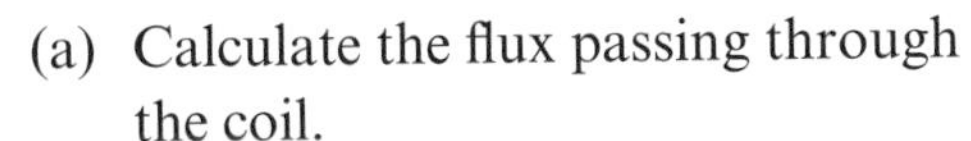

(a) Calculate the flux passing through the coil.

....

....

.... **(2 marks)**

(b) Calculate the flux linkage through the coil.

....

.... **(1 mark)**

4 A square coil of area A and N turns is placed into a uniform horizontal magnetic field of strength B. The coil is then rotated at a steady rate about a vertical axis as shown in the diagram.

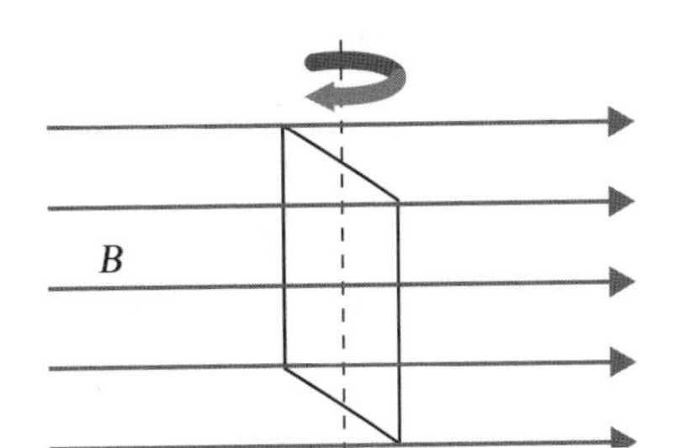

Sketch on the axes how the flux linkage through the coil varies during one rotation. Assume that the flux linkage is zero at $t = 0$ and increasing.

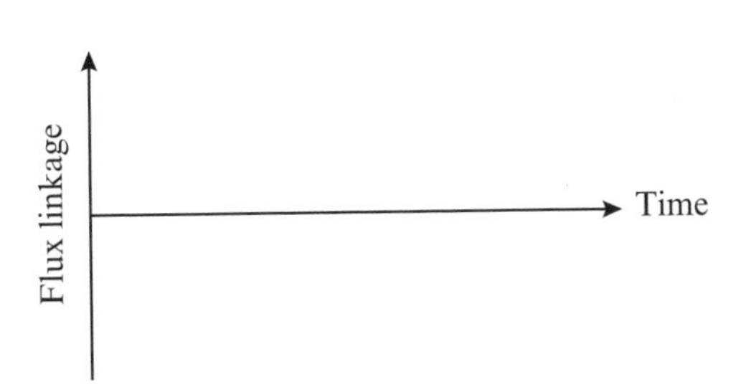

(3 marks)

Had a go ☐ Nearly there ☐ Nailed it! ☐

Faraday's law of electromagnetic induction and Lenz's law

1 A bar magnet is suspended from a spring above a coil as shown in the diagram.

With S open, the magnet is displaced and oscillates freely up and down. While it is oscillating, S is closed and the oscillations rapidly die away.

S
N
S

Guided

(a) Use Faraday's law to explain why there will be an induced e.m.f. but no induced current in the coil before the switch is closed.

- As the magnet moves up and down, its magnetic field creates a changing flux through the coil.
- Induced e.m.f. in the coil is .... flux linkage (Faraday's law).
- In this example, the switch .... , therefore there is ....

Use bullet points to make three separate, linked points.

(3 marks)

(b) Use Lenz's law to explain why the oscillations die away when the switch is closed.

....

....

....

(3 marks)

(c) Lenz's law is a consequence of the law of conservation of energy. Explain how this example illustrates the law of conservation of energy.

....

....

....

(2 marks)

Maths skills

2 A flat coil of 50 turns and area $25\,cm^2$ lies so that its plane is parallel to a uniform magnetic field of strength 20 mT. The coil is then rotated through 90° in a time of 0.20 s so that it finishes with its plane perpendicular to the field.

(a) Calculate the change in flux linkage through the coil as it is rotated through 90°.

....

....

(2 marks)

(b) Calculate the average induced e.m.f. in the coil during the rotation.

....

....

....

(3 marks)

The search coil

1 A search coil of 200 turns and cross-sectional area 4.0 cm^2 is placed so that its plane is perpendicular to an alternating magnetic field of peak value 20 mT and frequency 60 Hz. What is the peak value of the induced e.m.f. in the search coil?

☐ **A** 50 μV ☐ **B** 1.6 mV ☐ **C** 96 mV ☐ **D** 0.60 V **(1 mark)**

2 The diagram shows a simple a.c. generator consisting of a coil of 120 turns and area 5.0 cm^2 rotating in a uniform magnetic field of strength 40 mT. The coil rotates at a steady rate, completing one rotation in T seconds.

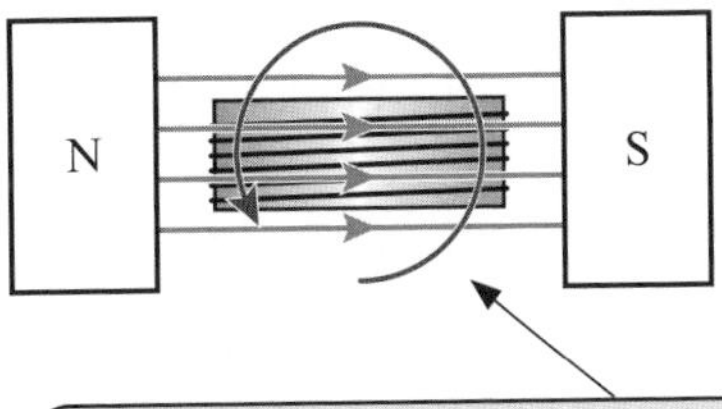

N S

Watch out for a change of sign of flux linkage.

(a) At $t = 0$ the coil is in the position shown and the flux linkage is increasing. Calculate the flux linkage through the coil at the following times:

(i) $t = 0$ ....

(ii) $t = 0.25\,T$ ....

(iii) $t = 0.50\,T$ ....

(iv) $t = 0.75\,T$ .... **(4 marks)**

(v) Hence sketch a graph on the axes below to show how the flux linkage varies with time during one complete rotation of the coil. Add a suitable scale and unit to the y-axis. **(4 marks)**

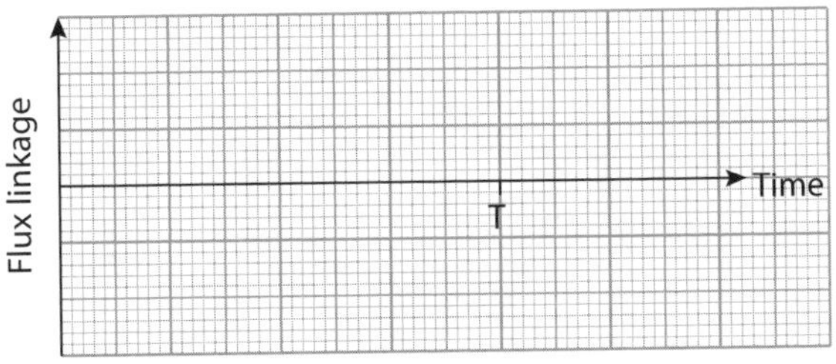

Flux linkage — Time — T

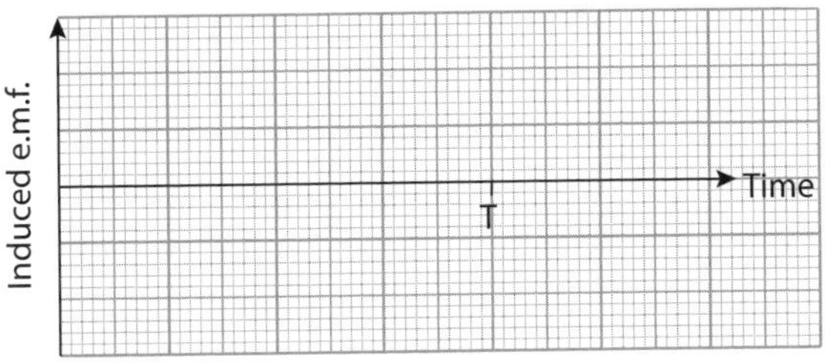

Induced e.m.f. — Time — T

(b) (i) Use the graph from part (a) to sketch a graph of how the induced e.m.f. in the coil varies during one rotation. Do not put a scale on the y-axis. **(2 marks)**

Guided

(ii) Explain the relationship between the graph of e.m.f. (in (b)) and the graph of flux linkage (in (a)).

.... law says that .... is proportional to negative rate of change of .... so the values on the e.m.f. graph are the .... of the flux-linkage graph at the same time.

If you are asked to 'explain' something, try to relate your answer to the underlying physical principles – in this case Faraday's law.

(3 marks)

Had a go ☐ Nearly there ☐ Nailed it! ☐

The ideal transformer

1 The diagram shows an ideal transformer with its primary coil connected to an a.c. supply of peak value 120 V and its secondary coil connected to a resistor of resistance 60 Ω.

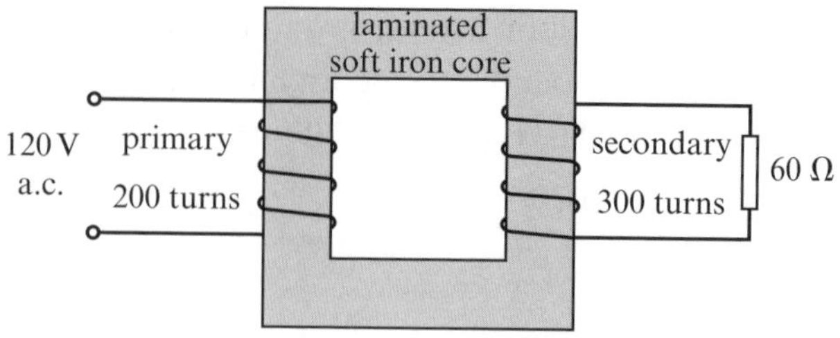

laminated soft iron core
120 V a.c.
primary 200 turns
secondary 300 turns
60 Ω

(a) Calculate the peak voltage across the resistor.

....

.... **(2 marks)**

(b) Calculate the peak current in the primary coil.

....

.... **(3 marks)**

(c) Ideal transformers are 100% efficient but real transformers are not. Explain why the core of a real transformer is laminated.

....

....

.... **(3 marks)**

Guided 2 The diagram shows an experimental set-up similar to one used by Michael Faraday when he investigated electromagnetic induction. When the switch is closed the compass needle deflects to the left and then returns to its original position. Explain, in as much detail as you can, why the needle deflects and returns.

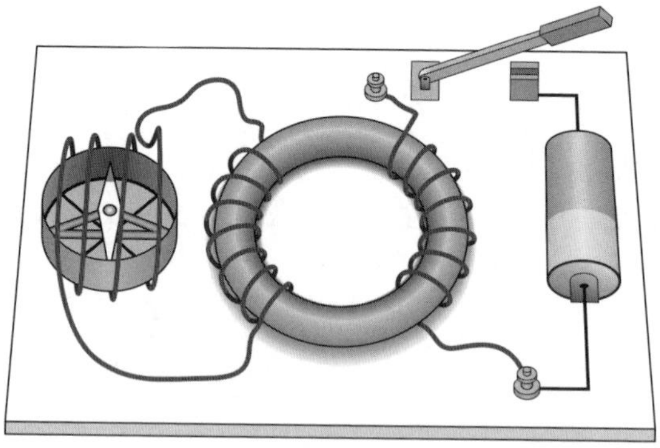

When the switch is closed the current in the primary coil increases from zero. This causes an increase in the magnetic flux in the iron core. This changing flux links the secondary coil and induces ....

....

....

....

....

....

....

....

.... **(8 marks)**

Exam skills

1 The graph shows how the current through a coil varies with time.

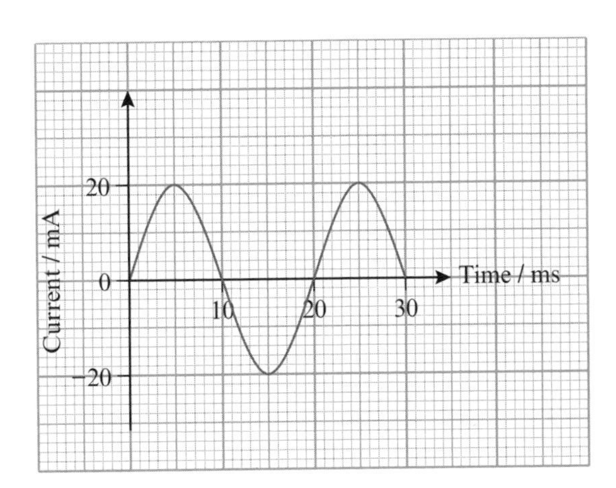

(a) Calculate the frequency of the a.c. current.

....

.... **(2 marks)**

A second coil is placed close to the first one, as shown in the diagram. Its terminals are X and Y. When an oscilloscope is connected to terminals X and Y, the trace on the screen shows an alternating e.m.f. between the terminals.

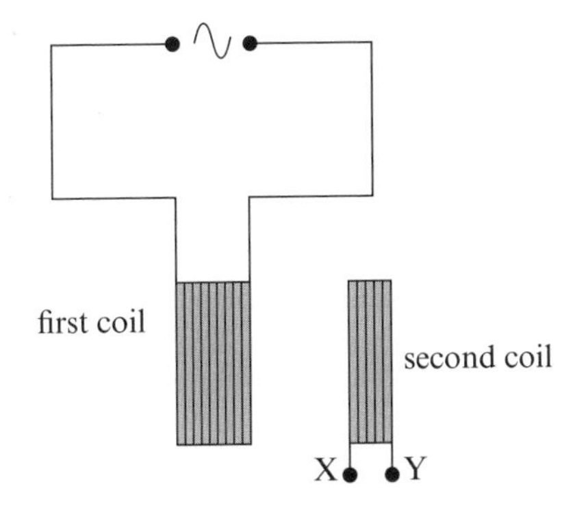

(b) Explain why an induced e.m.f. is produced in the second coil.

....

....

....

....

....

.... **(4 marks)**

(c) At what times will the e.m.f. in the second coil reach its peak values? Explain your answer.

....

....

....

....

.... **(3 marks)**

(d) Discuss whether there is any energy transfer between the two coils.

....

....

....

.... **(2 marks)**

Had a go ☐ Nearly there ☐ Nailed it! ☐

The nuclear atom

1 Rutherford's alpha-particle scattering experiment provided evidence for the nuclear model of the atom.

(a) Why was the experiment carried out in a vacuum?

....

.... **(1 mark)**

(b) Why was gold foil used as a target?

....

.... **(1 mark)**

Guided

(c) Which piece of evidence suggested that the atom is mainly empty space?

Most of the alpha particles passed through the gold foil with little

....

This suggests that they did not get close to ....

so most of the space inside the atom must be .... **(2 marks)**

(d) Explain how the scattering of alpha particles suggested that:

(i) the nucleus is charged ....

....

.... **(2 marks)**

(ii) the nucleus is tiny compared with the atom. ....

....

.... **(2 marks)**

(e) Complete the diagram below to show the paths of the three alpha particles as they approach the gold nucleus.

alpha particles

gold nucleus

Think about the volume of the atom and the mass of the particles in it according to the Rutherford model.

(3 marks)

(f) What does the Rutherford model suggest about the density of nuclear matter compared with the density of normal matter?

....

.... **(2 marks)**

(g) Gold can be represented symbolically by $^{197}_{79}\text{Au}$. How many protons and neutrons are present in the gold nucleus?

....

.... **(2 marks)**

Nuclear forces

1 The table below shows the nuclear radii r of five different elements in femtometres (1 fm = 10^{-15} m). A is the mass number.

Element	r / fm	A	log (r / fm)	log (A)
C	2.66	12		
Si	3.43	28		
Fe	4.35	56		
Sn	5.49	120		
Pb	6.66	208		

(a) Complete the table by calculating values for log (r/fm) and log(A). **(3 marks)**

(b) It is suggested that nuclear radius r varies with mass number A according to the relationship $r = r_0A^n$. Explain how a log–log graph can be used to test this relationship.

....

....

....

.... **(3 marks)**

(c) Use the grid below to plot such a graph and determine the values of r_0 and n.

(5 marks)

Had a go ☐ Nearly there ☐ Nailed it! ☐

The Standard Model

1 Which of the following includes all of the particles found in a $^{4}_{2}He$ atom?

☐ **A** four baryons and two mesons

☐ **B** two baryons and two leptons

☐ **C** four hadrons and two leptons

☐ **D** four leptons and two baryons. **(1 mark)**

2 Which type of force binds the nucleus of the helium atom to the orbiting electrons?

☐ **A** gravitational force

☐ **B** weak nuclear force

☐ **C** electromagnetic force

☐ **D** strong nuclear force. **(1 mark)**

3 Which of the following is regarded as a fundamental particle in the Standard Model?

☐ **A** the top quark

☐ **B** the hydrogen nucleus

☐ **C** the neutron

☐ **D** the pi-meson. **(1 mark)**

4 An anti-deuterium atom would consist of a nucleus of one antiproton and one antineutron with a single orbiting antielectron.

(a) What force binds the antiproton and antineutron together?

.... **(1 mark)**

(b) State the quark structure of

(i) the antineutron: ....

(ii) the antiproton: .... **(2 marks)**

(c) What are the similarities and differences between the antielectron and the electron?

....

....

.... **(3 marks)**

(d) Your body is made from just three fundamental particles. Name these three fundamental particles.

....

....

.... **(3 marks)**

The quark model of hadrons

1 Two different types of baryon called omega particles have quark structures: uds and sss. Which line in the table below gives the correct charges for these baryons?

	Quark structure uds	Quark structure sss
☐ A	0	e
☐ B	0	$-e$
☐ C	$-e$	$2e$
☐ D	e	$-2e$

(1 mark)

Guided **2** Free neutrons are unstable particles and decay spontaneously to protons.

(a) Show that this decay must involve a change of quark flavour.

Both neutrons and protons are baryons consisting of three quarks.

Neutrons are udd and protons ...., so when a neutron decays

to a proton a .... quark must change to .... **(2 marks)**

(b) Neutrons and protons both have baryon number +1 but they have different charges. Suggest what must happen when a neutron decays in order to conserve charge.

.... **(2 marks)**

3 Particle physicists often carry out experiments in which proton beams collide a target containing more protons (e.g. liquid hydrogen). This often results in creation of pions. Here are two possible reactions to create pions:

$^{1}_{1}p + ^{1}_{1}p \rightarrow ^{1}_{1}p + ^{1}_{1}p + ^{0}_{0}\pi^{0}$ $\qquad$ $^{1}_{1}p + ^{1}_{1}p \rightarrow ^{1}_{1}p + ^{1}_{0}n + ^{0}_{1}\pi$

(a) State:

(i) the type of particle the pions are

....

(ii) the baryon number of a pion

....

(iii) the structure of a π^{0} particle

....

(iv) the structure of a π^{+} particle

.... **(4 marks)**

(b) Explain how it is possible to start off with two protons and end up with two protons **and** a pion in the first reaction without violating the law of conservation of energy.

....

....

.... **(2 marks)**

Had a go ☐ Nearly there ☐ Nailed it! ☐

β^- decay and β^+ in the quark model

1 Which of the statements below is **not** correct about beta decay?

☐ **A** Beta-plus decay conserves baryon number.

☐ **B** Beta-plus decay involves the emission of a neutrino.

☐ **C** Beta-minus decay conserves charge, energy, momentum and lepton number.

☐ **D** Beta-minus decay occurs in proton-rich nuclei. **(1 mark)**

2 Carbon-14 is unstable and decays by beta-minus decay. Here is an equation for the decay:

$^{14}_{6}C \rightarrow ^{14}_{7}N + ^{0}_{-1}e + ^{0}_{0}\bar{\nu}$

(a) Explain why the emission of a beta-particle also requires the emission of an antineutrino.

Think about conservation of lepton number and the fact that the anti-neutrino is an antiparticle.

....

....

....

.... **(2 marks)**

(b) How does the nitrogen nucleus differ from the carbon nucleus in this decay?

....

.... **(2 marks)**

(c) It is suggested that the underlying decay is actually: $^{1}_{0}n \rightarrow ^{1}_{1}p + ^{0}_{-1}e + ^{0}_{0}\bar{\nu}$

(i) Complete the table below by ticking the correct box to identify each particle by type:

	Quark	Lepton	Meson	Baryon
Proton				
Neutron				
Electron				
Neutrino				

(4 marks)

(ii) Explain how baryon number is conserved in beta decay.

....

.... **(1 mark)**

Guided

(iii) Explain how the decay above can be described in terms of a change of quark flavour.

A neutron consists of .... quarks. In beta-minus decay, a neutron decays to become a .... Since protons consist of .... quarks, ... quark must have changed to .... quark in the process. This is a change of flavour. **(2 marks)**

Radioactivity

Guided

1 (a) Explain why gamma rays have a much longer range in air than beta particles or alpha particles.

The range of an ionising radiation depends on how strongly it interacts with matter. Gamma rays are .... so they are only .... ionising and therefore they lose energy slowly as they move through the air. Alpha particles and beta particles are both .... ionising so they lose energy .... and have .... ranges. **(2 marks)**

(b) Explain why alpha sources are far more dangerous to humans if taken inside the body (e.g. inhaled radon gas) than if accidentally handled.

....

.... **(2 marks)**

2 A student is given a sample of a radioactive source to test. She uses a Geiger counter to measure count rates over 2-minute intervals with and without the source and with three different absorbers. Here are her results:

Source	Absorber	Counts (1)	Counts (2)	Counts (3)	Counts (4)	Counts (5)	Counts (avg)
None	none	22	20	20	25	18	21.0
Present	none	77	83	85	81	80	
Present	thick card	72	74	78	75	80	
Present	2 mm Al	25	22	17	19	20	20.6
Present	5 mm Pb	18	24	19	22	21	20.8

(a) Explain why repeat readings of the number of counts in any 2-minute interval vary.

....

.... **(1 mark)**

(b) Complete the table by including the missing average counts. **(2 marks)**

(c) Identify the most likely type of radiation emitted by this source and explain your reasoning.

Use a process of elimination – which types of radiation can be stopped by each type of absorber?

....

....

....

....

.... **(4 marks)**

Balancing nuclear transformation equations

1 The diagram shows a naturally occurring decay chain that starts at uranium-238 and ends with the stable isotope lead-206.

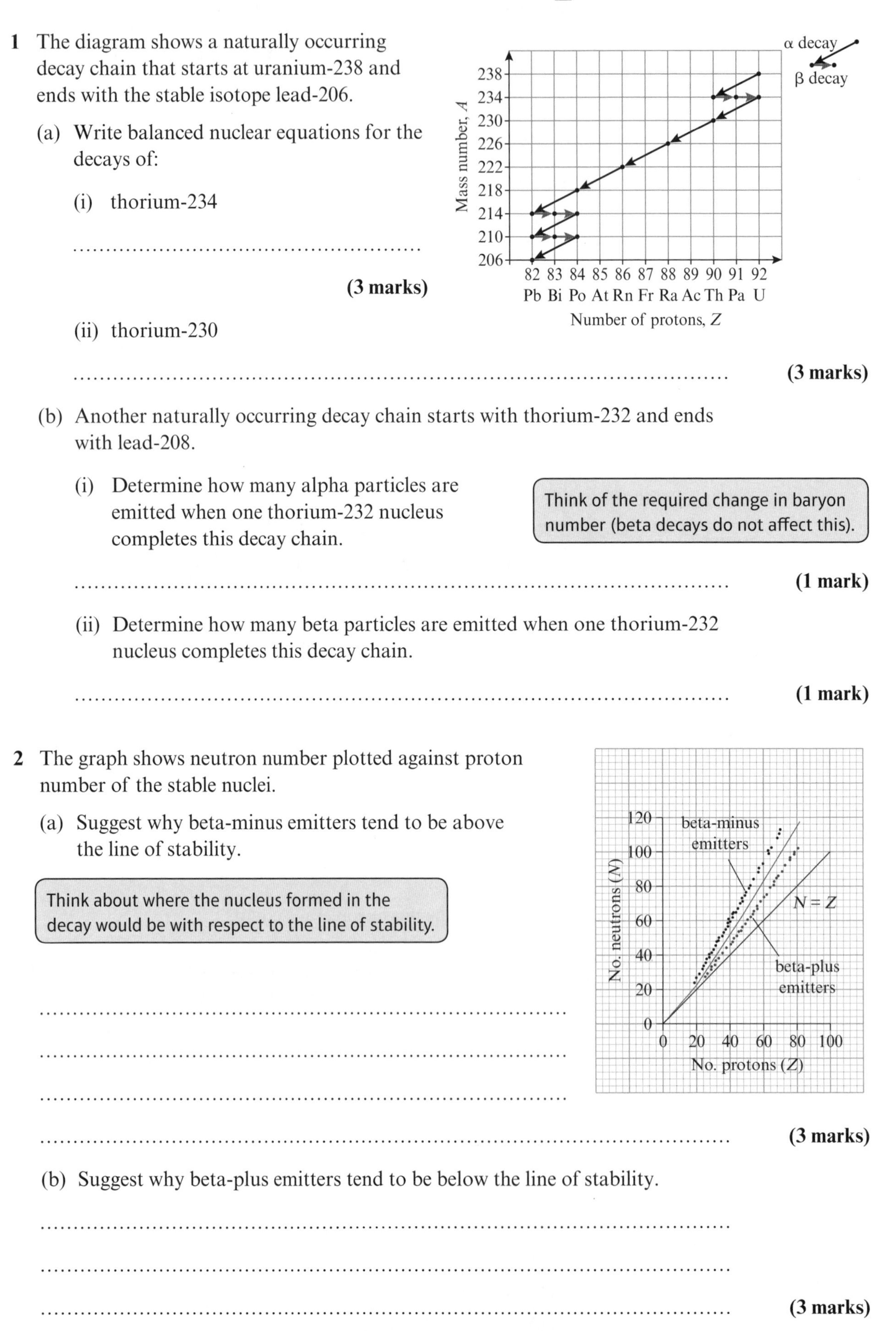

(a) Write balanced nuclear equations for the decays of:

(i) thorium-234

....

(3 marks)

(ii) thorium-230

.... **(3 marks)**

(b) Another naturally occurring decay chain starts with thorium-232 and ends with lead-208.

(i) Determine how many alpha particles are emitted when one thorium-232 nucleus completes this decay chain.

> Think of the required change in baryon number (beta decays do not affect this).

.... **(1 mark)**

(ii) Determine how many beta particles are emitted when one thorium-232 nucleus completes this decay chain.

.... **(1 mark)**

2 The graph shows neutron number plotted against proton number of the stable nuclei.

(a) Suggest why beta-minus emitters tend to be above the line of stability.

> Think about where the nucleus formed in the decay would be with respect to the line of stability.

....

....

....

.... **(3 marks)**

(b) Suggest why beta-plus emitters tend to be below the line of stability.

....

....

.... **(3 marks)**

Radioactive decay 1

1 Two radioactive sources, X and Y, have equal activity at a particular time. The half-life of X is 20 minutes and the half-life of Y is 30 minutes. Calculate the ratio of the activity of X to the activity of Y after 1 hour.

☐ **A** 1:2 ☐ **B** 2:3 ☐ **C** 3:2 ☐ **D** 2:1 **(1 mark)**

2 In an experiment to determine the half-life of a certain mass of a radioactive gas, an experimenter recorded the count rate per minute (cpm) against time for a period of 4 minutes. He then repeated the experiment with the same quantity of the same radioactive gas. His results are shown in the table.

Time / s	Trial 1 / cpm	Trial 2 / cpm	Mean / cpm
0	205	225	
30	155	163	
60	112	120	
90	83	93	
120	65	73	
150	50	58	
180	43	47	
210	36	40	
240	34	38	

(a) Complete the table by calculating the mean count rates. **(2 marks)**

(b) Use the graph paper below to plot a graph of mean count rate against time.

(3 marks)

(c) Use your graph to estimate the approximate average background count rate.

.... **(1 mark)**

(d) Use the graph to determine a value for the half-life of this radioactive gas. Show your working on the graph.

Don't forget to take account of the average background count.

half-life = ... **(3 marks)**

Radioactive decay 2

1 Cobalt-60 is produced in nuclear reactors. A laboratory uses a small sample of cobalt-60 as a gamma-ray source. The activity of the source when supplied new was 186 kBq, and the half-life of cobalt-60 is 5.3 years. The laboratory acquires a newly produced cobalt-60 source of the same mass.

Guided

(a) A student uses a Geiger–Müller tube to measure the intensity of gamma radiation from the newly acquired cobalt-60 source and finds that the count rate is very much less than 186 000 counts per second. Suggest three reasons for this.

The gamma-rays are emitted randomly in all directions so ....

....

The efficiency of the detector is ....

....

The body of the sample itself ....

.... **(3 marks)**

(b) Calculate the activity of the old source:

(i) 5.3 years after purchase

.... **(1 mark)**

(ii) 15.9 years after purchase.

.... **(2 marks)**

(c) (i) Calculate the decay constant for cobalt-60 and state its unit.

....

....

.... **(1 mark)**

(ii) The source must be replaced when its activity falls below 10% of its initial activity. How often must cobalt-60 sources be replaced?

....

....

....

....

.... **(3 marks)**

Einstein's mass–energy equation

1 It is usually assumed that mass is conserved in chemical reactions even if energy is released. For example, when hydrogen is burnt in oxygen the energy released is $1.43 \times 10^8\ \mathrm{J\,kg^{-1}}$.

Maths skills

(a) (i) Calculate the mass equivalent to $1.43 \times 10^8\ \mathrm{J}$ of energy.

....

.... **(2 marks)**

(ii) Hence explain why it is sensible to assume that chemical reactions conserve mass.

....

....

.... **(2 marks)**

(b) (i) Hydrogen is converted to helium by nuclear fusion reactions in the core of the Sun. The energy released is about $6.8 \times 10^{14}\ \mathrm{J\,kg^{-1}}$ for the helium produced. Use this value to calculate the percentage change in mass in this nuclear fusion reaction.

....

....

.... **(2 marks)**

(ii) State why it is important to consider mass changes in nuclear reactions.

....

.... **(1 mark)**

(c) The most efficient way to release energy from matter is to combine matter and antimatter, when all mass is converted into energy. Using your answers to (a)(i) and (b)(i) above, calculate how many times greater the energy released per kilogram in matter–antimatter annihilation is than:

(i) a chemical reaction such as the combustion of hydrogen

....

....

....

(ii) a nuclear fusion reaction such as the creation of helium in the Sun's core.

....

....

.... **(4 marks)**

Had a go ☐ Nearly there ☐ Nailed it! ☐

Binding energy and binding energy per nucleon

1 Which of the following statements gives the most accurate definition of the binding energy of an atomic nucleus?

☐ **A** the energy that binds the atom together

☐ **B** the energy stored in the nucleus

☐ **C** the energy needed to take the nucleus apart

☐ **D** the energy that binds the nucleons together. **(1 mark)**

Guided

2 Calculate the binding energy per nucleon for a $^{16}_{8}O$ oxygen nucleus. Express your answer in J and eV. (Mass of an oxygen nucleus 15.994915 u; m_p = 1.007276 u; m_n = 1.008665 u; 1 u = 1.67×10^{-27} kg.)

Work out the mass deficit for the nucleus compared with the nucleons and then convert this to energy. Don't forget to divide by the number of nucleons.

mass of 8 protons and 8 neutrons = ...

mass deficit of nucleus Δm = ...

total binding energy = $c^2\Delta m$ = ...

....

binding energy per nucleon = ...

.... **(4 marks)**

3 The diagram shows how the binding energy per nucleon varies with mass number.

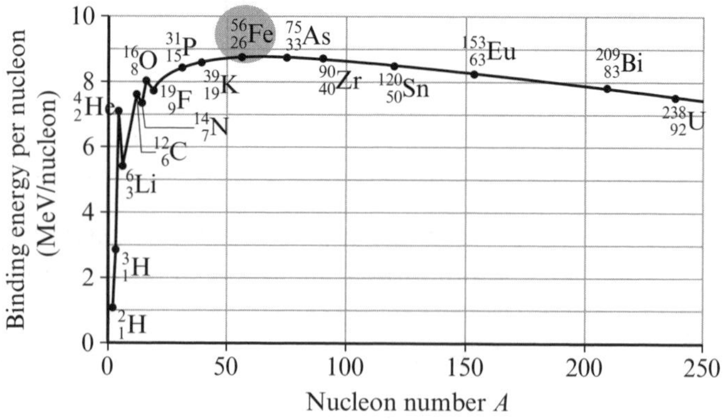

(a) Explain why iron-56 is regarded as the most stable nuclide.

....

.... **(2 marks)**

(b) After hydrogen, the next three most abundant nuclides in the Milky Way galaxy are helium (He), oxygen (O) and carbon (C). Use the graph to explain why this might be.

....

....

.... **(3 marks)**

(c) Estimate the energy that could be released if a carbon-12 nucleus could be formed by fusing three helium-4 nuclei.

....

....

.... **(2 marks)**

Nuclear fission

1 The used fuel rods from nuclear fission contain daughter nuclei that may be highly unstable radioisotopes. Which statement about these fuel rods is **not** correct?

☐ **A** They contain both short- and long-lived radioactive isotopes.

☐ **B** They must be treated as intermediate-level nuclear waste.

☐ **C** They require cooling for several years.

☐ **D** They remain radioactive for several thousand years. **(1 mark)**

2 (a) In the space below, sketch a graph of nuclear binding energy against nucleon number.

(3 marks)

(b) Explain, with reference to your diagram from part (a), how nuclear fission of heavy nuclei can release a large amount of energy.

> You might find it helpful to add lines or labels to your graph.

....

....

.... **(3 marks)**

3 Here is an incomplete nuclear equation for the fission of a uranium-235 nucleus after it absorbs a neutron. Write the missing values below.

$^{a}_{b}\text{n} + ^{235}_{92}\text{U} \rightarrow ^{c}_{d}\text{Ba} + ^{92}_{36}\text{Kr} + 3^{a}_{b}\text{n}$

a = ... b = ... c = ... d = ... **(4 marks)**

4 The energy released by nuclear fission of uranium-235 is about 200 MeV per fission.

Calculate the energy released if all of the atoms in 1.0 g of uranium-235 were to undergo nuclear fission.

> First calculate the number of atoms using the Avogadro number (6.02×10^{23}). Then multiply by the energy per fission in joules.

....

....

....

.... **(3 marks)**

Had a go ☐ Nearly there ☐ Nailed it! ☐

Nuclear fusion and nuclear waste

Guided

1 (a) Describe the process of nuclear fusion and explain why it is hard to achieve in a controlled way.

Two nuclei have to approach one another close enough for the .... force to overcome the .... repulsion (since both nuclei are positively charged). The .... nuclear force can then bind them together to form a heavier nucleus with the release of .... This is very difficult to achieve in practice because the individual nuclei must have extremely high .... to get so close. This means creating and controlling plasmas at extremely .... **(4 marks)**

(b) Two deuterium nuclei can fuse to form a tritium nucleus plus one proton. The equation for such a reaction is:

$^{2}_{1}H + ^{2}_{1}H \rightarrow ^{3}_{1}H + ^{1}_{1}H$

(i) Show how this reaction conserves both charge and baryon number.

....

.... **(2 marks)**

(ii) Calculate the energy released by this reaction. Express your answer in J and MeV.

Mass of a proton = 1.007276 u; mass of a deuterium nucleus = 2.013553 u; mass of a tritium nucleus = 3.015500 u; 1 u = 1.67×10^{-27} kg.

....

....

....

....

Start by working out the mass deficit for the reaction, then use $E = mc^2$ and finally convert units.

.... **(5 marks)**

(iii) Deuterium occurs naturally at a ratio of about 1 in 4500 hydrogen atoms. Suppose that all of the deuterium within 1.0 kg of seawater were to undergo nuclear fusion to form tritium by the reaction above. Use the ratio above and your answer to (c)(ii) to estimate the maximum energy that could be released from the seawater. Assume that the average molar mass of the water (H_2O) is 18 g, and dissolved salts can be neglected. Avogadro number = 6.02×10^{23}.

....

....

....

....

Start by calculating the number of molecules of water and then find the number of deuterium atoms.

.... **(3 marks)**

Exam skills

1 A mass spectrometer is an instrument used to separate beams of charged particles by their mass. It has two main parts, a velocity selector, which ensures that all of the charged particles are travelling at the same velocity, and a region of uniform magnetic field, in which the particles are deflected. The diagram shows the principle of the velocity selector.

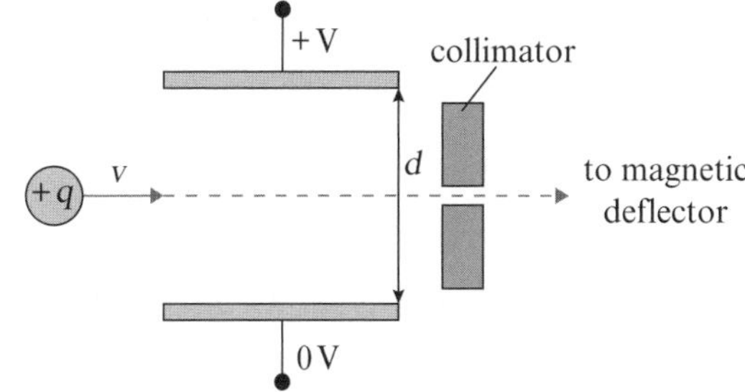

(a) A particle with charge $+q$ moving horizontally with velocity v enters the region between the two plates. Write down an expression for the magnitude of the force from the electric field acting on the particle and state its direction.

.... **(2 marks)**

(b) The charged particle continues to move horizontally as it passes between the plates.

(i) State the direction of the magnetic field.

.... **(1 mark)**

(ii) Derive an expression for the magnetic field strength $B1$.

....

.... **(2 marks)**

(iii) Explain why particles travelling faster or slower than v will not reach the magnetic deflector.

....

....

.... **(2 marks)**

(c) The diagram shows a beam of charged particles after it leaves the velocity selector and as it enters the region of uniform magnetic field. The beam contains two isotopes, P of mass m_1, and Q of mass m_2, where $m_2 > m_1$.

X
Y
+q
region of magnetic field of strength into page

The charged particles are deflected in arcs of circles as shown.

(i) Which isotope collects at X? Explain your answer.

....

.... **(2 marks)**

(ii) Derive an expression for the distance XY.

....

....

.... **(3 marks)**

Production of X-ray photons

1 Which of the following statements about X-rays is **not** correct?

☐ **A** They are transverse waves.
☐ **C** They are not absorbed by matter.
☐ **B** They are ionising.
☐ **D** They cannot be polarised. **(1 mark)**

2 The diagram shows an X-ray tube.

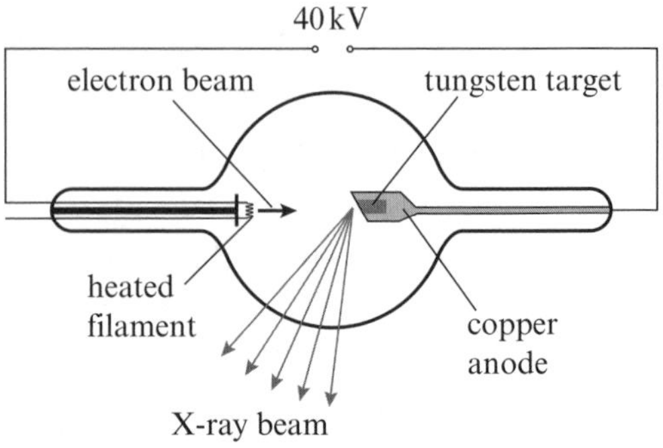

40 kV
electron beam
tungsten target
heated filament
copper anode
X-ray beam

(a) Explain why:

(i) the electrons must be accelerated through a large potential difference in order to produce X-rays.

....

.... **(2 marks)**

(ii) most of the anode is made from copper, and some anodes have water pumped through them.

....

.... **(2 marks)**

(iii) there is a vacuum in the tube.

....

.... **(1 mark)**

(b) Calculate the shortest wavelength of X-ray photons that can be emitted from this X-ray tube.

....

....

.... **(3 marks)**

Guided

3 The diagram shows a typical X-ray spectrum.

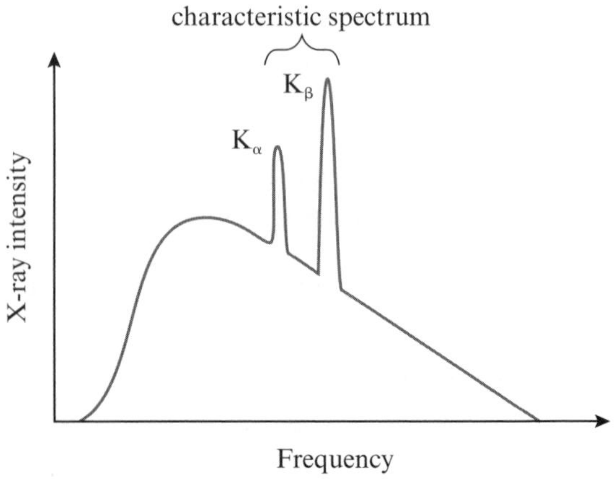

characteristic spectrum
K_β
K_α
X-ray intensity
Frequency

Explain how the two peaks K_α and K_β are produced. Your answer should explain why K_α has a shorter wavelength than K_β.

Incoming electrons eject atomic electrons in the target from .... energy levels. These lines are formed when outer .... in target atoms make quantum jumps down into the lower .... and emit a .... of equivalent energy (and thus defined wavelength). The K_α line has shorter wavelength so corresponds to a .... energy jump than the K_β line. **(3 marks)**

X-ray attenuation mechanisms

1 Which of the following mechanisms is only possible for extremely short wavelength X-rays?

☐ **A** the photoelectric effect

☐ **B** pair production

☐ **C** Compton scattering

☐ **D** simple scattering

(1 mark)

Guided

2 When X-rays are absorbed by a particular material their intensity falls to 50% after travelling a distance of 4.5 cm in the material.

(a) What percentage of the original intensity emerges when these X-rays pass through a slab of the same material of thickness 18 cm?

Since the intensity decays exponentially it will have a constant half .... so the intensity halves every .... cm. 18 cm corresponds to four .... thicknesses so the intensity will be reduced to $\frac{1}{2}n$ where $n =$ .... The transmitted intensity is therefore ....

(2 marks)

(b) Calculate a value for the attenuation (absorption) coefficient of this material for these X-rays.

Maths skills: To solve an equation with an exponential term, take logs.

....

....

....

(3 marks)

(c) What thickness of this material would be required to reduce the intensity to 1% of its original value?

....

....

(3 marks)

3 (a) Calculate the minimum photon energy needed if an X-ray photon is to create an electron–positron pair when it interacts with a nucleus. The mass of an electron is $m_e = 9.1 \times 10^{-31}$ kg.

....

....

....

(2 marks)

(b) Would it be possible for an X-ray of wavelength 1.0×10^{-12} m to produce an electron positron pair in this way? Support your answer with a calculation.

A shorter wavelength corresponds to a higher photon energy.

....

....

....

(3 marks)

Had a go ☐ Nearly there ☐ Nailed it! ☐

X-ray imaging and CAT scanning

1 Which of the following statements about X-ray imaging is correct?

☐ **A** A contrast medium must have a low attenuation coefficient.

☐ **B** The higher the frequency of the X-rays used, the lower the attenuation.

☐ **C** Materials with higher atomic number transmit X-rays more effectively.

☐ **D** Higher density materials have larger half-thicknesses for X-ray absorption. **(1 mark)**

2 The graph shows how the attenuation of X-rays varies with photon energy for fat, muscle and is correct?

Absoroption coefficient
iodine
fat
muscle
X-ray photon energy

☐ **A** If iodine is introduced into blood vessels, they will stand out against muscle in an X-ray image.

☐ **B** Muscles can be clearly distinguished from fat in X-ray images.

☐ **C** Iodine has a lower absorption coefficient than fat at the same photon energy.

☐ **D** X-rays pass through iodine more easily than through fat or muscle. **(1 mark)**

Guided **3** The image shows an X-ray image of a hand.

(a) Explain how you can tell that this is a negative image.

The brightest parts of the image are the ....

These actually .... more of the X-rays than

the ...., so if the image brightness were

proportional to the X-rays transmitted, then these

areas would be darkest. **(2 marks)**

(b) Explain why there is good contrast between the images of the bones and the soft tissues surrounding them.

....

.... **(2 marks)**

4 List the advantages and disadvantages of CAT scans compared to conventional X-ray techniques that use contrast media such as barium.

....

....

....

....

.... **(4 marks)**

Medical tracers

1 Which of the following radioisotopes would be most useful as a tracer in medicine?

☐ **A** a gamma-emitting radioisotope with half-life 8 months

☐ **B** an alpha-emitting radioisotope with half-life 8 hours

☐ **C** a gamma-emitting radioisotope with half-life 8 hours

☐ **D** an alpha-emitting radioisotope with half-life 8 months. **(1 mark)**

2 Technetium-99m ($^{99}_{43}$Tc) is produced from the beta-minus decay of molybdenum-99 ($^{99}_{42}$Mo). The half-life of molybdenum-99 is 66 hours and the half-life of technetium-99 is 6 hours.

(a) Write down a balanced nuclear decay equation to show how technetium-99 is created by the decay of molybdenum-99.

.... **(2 marks)**

Guided

(b) Molybdenum is usually created in a nuclear reactor and then transported to hospitals where technetium is extracted for use as a tracer. Explain why the relatively long half-life of the molybdenum allows it to be produced at a distance and then transported.

The relatively long .... means that only ....

.... has decayed during transportation so the technetium generator is still useful when it arrives at the hospital. **(2 marks)**

(c) Technetium-99m is a metastable state of technetium-99 that decays by emitting a gamma ray. The half-life for this decay is 6 hours. Explain why technetium-99m is ideal for use as a radioactive tracer in nuclear medicine.

....

....

.... **(3 marks)**

3 (a) Beta-plus emitters such as fluorine-18 are often used in nuclear medicine. The radiation detected is not beta-plus particles, it is pairs of gamma-rays. Explain how the gamma rays are produced.

....

....

.... **(3 marks)**

(b) Fluorine-18 has a half-life of just 110 minutes. State one advantage and one disadvantage of having such a short half-life.

....

....

.... **(2 marks)**

Had a go ☐ Nearly there ☐ Nailed it! ☐

The gamma camera and diagnosis

1 What is the purpose of a gamma camera?

☐ **A** to form images of structures inside the body

☐ **B** to kill tumours

☐ **C** to monitor biological processes inside the body

☐ **D** to identify bone damage. **(1 mark)**

2 A gamma camera consists of a collimator, a scintillator, photomultiplier tubes, circuitry and a computer to produce a display image. Explain briefly what each of the following parts does.

(a) The scintillator

....

....

....

(b) The photomultiplier tubes

....

....

.... **(4 marks)**

3 The diagram shows part of a photomultiplier tube.

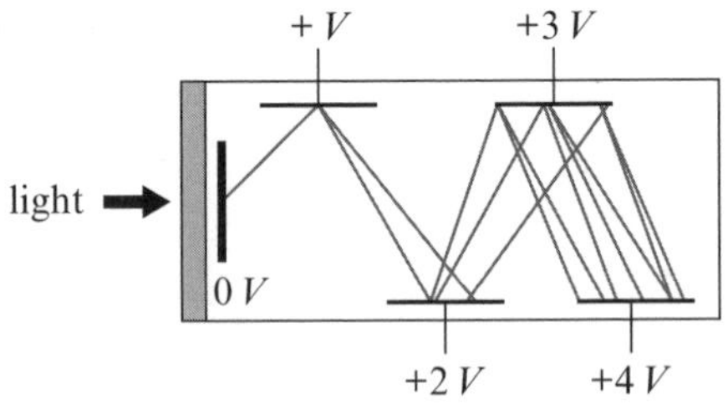

(a) Explain why the voltage of successive electrodes becomes increasingly positive.

....

....

.... **(2 marks)**

Guided

(b) In a particular photomultiplier there are 12 stages of amplification and each stage increases the electron current by a factor of 5. What is the charge created by the photomultiplier for each photon striking it? (Charge on an electron = 1.6×10^{-19} C.)

Each photon ejects one electron from the photocathode. This is multiplied five times at each of 12 anodes so the final number of electrons is .... to the power .... and the total charge is .... $\times 1.6 \times 10^{-19}$ C = ... **(3 marks)**

PET scanning and diagnosis

1 Which of the following statements about PET scanning is **not** correct?

☐ **A** PET scanning involves the annihilation of matter and antimatter inside the body.

☐ **B** PET scanning involves gamma emission from a radioisotope.

☐ **C** PET scanning involves a beta-plus decay.

☐ **D** PET scanning can be used to image brain functions. **(1 mark)**

2 The diagram shows how ring detectors are used to detect gamma rays.

gamma camera

(a) Explain why signals are only recorded when two detectors on opposite sides of the ring are triggered almost simultaneously.

....

....

.... **(2 marks)**

Guided

(b) Explain how small differences in the time of arrival of a pair of gamma-ray photons can be used to calculate the position at which the gamma rays were emitted.

If an event occurs closer to one side of the ring ....

The difference in .... can be used to calculate the difference

in .... by each .... so it is possible to calculate where

the original event was along a straight line running from ....

on the ring to the one opposite it. **(2 marks)**

(c) In a particular PET scanner, the diameter of the ring detector is 2.0 m and the patient is in the centre of the ring. A pair of photons from the same decay inside the patient's brain arrive at detectors on opposite sides of the ring with a time difference of 200 ps (1 ps = 10^{-12} s). Assume the paths of the two photons lie along the same diameter of the ring.

(i) How far from the centre of the ring was the decay?

....

....

.... **(3 marks)**

(ii) Explain why the speed of response of the detectors will affect the resolution of the image (i.e. the amount of detail that can be seen in the image).

....

....

.... **(3 marks)**

Had a go ☐ Nearly there ☐ Nailed it! ☐

Ultrasound

1 Ultrasound is sound with a frequency greater than 20 000 Hz. Which of the following wavelengths corresponds to ultrasound waves in air but not in water? (The speed of ultrasound in air is $340\,m\,s^{-1}$ and in water is $1500\,m\,s^{-1}$.)

☐ **A** 5 m ☐ **B** 50 cm ☐ **C** 5 cm ☐ **D** 5 mm **(1 mark)**

2 In a medical examination by ultrasound, a coupling gel is placed between the ultrasound transmitter/detector and the patient's skin. This will:

☐ **A** increase the intensity of the ultrasound signal

☐ **B** increase the proportion of ultrasound transmitted into the patient

☐ **C** decrease reflections from tissue boundaries inside the patient

☐ **D** decrease the time for the pulses to reflect and return. **(1 mark)**

3 Medical ultrasound imaging typically uses ultrasound frequencies between 2.0 and 10 MHz. The smallest detail that can be imaged is related to the wavelength of the waves used to form the image. The speed of ultrasound in tissue is $1400\,m\,s^{-1}$.

(a) Calculate the time between emitting a pulse of ultrasound and detecting the reflected ultrasound pulse if the distance from the transmitter to the reflecting boundary is 4.8 cm.

....

....

.... **(2 marks)**

Guided

(b) Suggest, with reasons, a maximum duration for the ultrasound pulses used in this scanner. Assume that images are formed from boundaries between 2.0 cm and 12 cm from the transmitter.

The shortest time between emitting a pulse and detecting its echo is for boundaries at ....

The time for the return trip of a pulse to this boundary is ....

In order that the returning pulse cannot be confused with the emitted pulse, the maximum pulse duration must be .... **(3 marks)**

(c) Discuss whether details 0.50 mm across can be imaged using ultrasound in the range 2.0–10 MHz.

....

....

....

....

.... **(4 marks)**

Acoustic impedance and the Doppler effect

1 Which of the following shows the correct SI units for acoustic impedance?

☐ **A** $kg\,m^{-2}\,s^{-1}$ ☐ **B** $kg\,m^{-4}\,s^{-1}$ ☐ **C** $kg\,m^{-2}\,s$ ☐ **D** $kg\,m^{-4}\,s$ **(1 mark)**

2 (a) Explain what is meant by acoustic impedance.

....

.... **(1 mark)**

(b) Explain how the acoustic impedances of two materials determines the percentage of ultrasound that is reflected and transmitted at a boundary between these two materials.

....

....

.... **(2 marks)**

(c) Complete the table by calculating the missing values.

Medium	Speed / $m\,s^{-1}$	Density / $kg\,m^{-3}$	Z / $kg\,m^{-2}\,s^{-1}$
Air	330	0.0012	
Water	1480	1000	1.5×10^6
Bone	4080		7.8×10^6
Muscle		1070	1.7×10^6
Blood	1550	1060	

(4 marks)

(d) Use values from the table to calculate the percentage of ultrasound that is reflected at a boundary between muscle and bone.

....

.... **(2 marks)**

3 The diagram shows a schematic of the arrangement used to measure blood flow rates.

Guided

Calculate the maximum percentage change in the ultrasound frequency that could be caused by blood cells moving at $40\,cm\,s^{-1}$. The speed of ultrasound in tissue is about $1500\,m\,s^{-1}$.

The fractional change in frequency is given by:

$$\frac{\Delta f}{f} = 2 \times \frac{v\cos\theta}{c}$$

This will be a maximum when $\cos\theta =$ ....

The percentage change is $\frac{\Delta f}{f} \times 100 =$ ....

.... **(3 marks)**

Ultrasound A-scan and B-scan

1 Which of the following statements about ultrasound scans is **not** correct?

☐ **A** B-scans provide information about the brightness of the object being scanned.

☐ **B** A-scans can be used to measure the distance to tissue boundaries.

☐ **C** Only B-scans produce a recognisable image of the object being scanned.

☐ **D** In A and B scans, A stands for amplitude and B stands for brightness. **(1 mark)**

2 The diagram shows two reflected pulses produced by an A-scan. Each horizontal square represents a time of 25 μs.

(a) Label the diagram to indicate the pulse reflected from the front surface and the pulse reflected from the back surface of the object being measured. **(2 marks)**

(b) Calculate the depth of the object.
(Ultrasound travels at $1500\,m\,s^{-1}$ in living tissue.)

....

....

.... **(3 marks)**

Guided (c) Estimate, with reasons, the uncertainty in your answer to (b).

The spread of the pulses is an indication of uncertainty in

....

Here, this uncertainty is about .... μs.

This corresponds to an uncertainty in total distance of

.... and an uncertainty in the measurement of about

.... **(3 marks)**

(d) State two reasons why the second pulse has a smaller amplitude than the first pulse.

....

....

.... **(2 marks)**

(e) Explain how a B-scan differs from an A-scan.

....

....

....

.... **(3 marks)**

Answers

1. Quantities and units

1 C – charge is not an SI base unit **(1)**

2 D – work is force × distance therefore units are $kg\,m\,s^{-2} \times m = kg\,m^2\,s^{-2}$ **(1)**

3 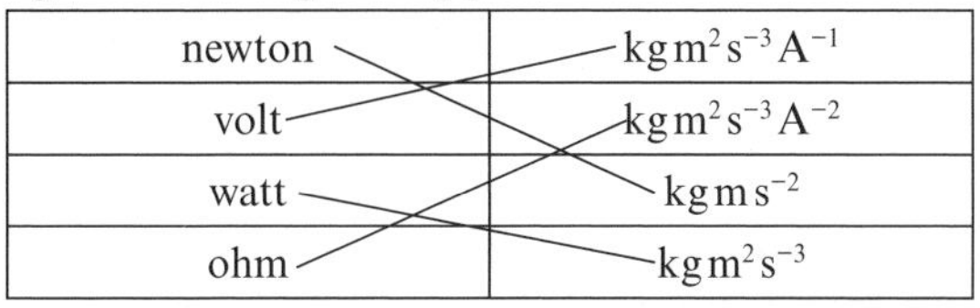

(3)

4 (a) Rearranging the equation gives $\rho = \frac{RA}{l}$. Considering the units of the quantities on the right-hand side gives $\frac{\Omega\,m^2}{m} = \Omega\,m$ **(2)**

(b) $R = 1.68 \times 10^{-8} \times \frac{1500}{(5 \times 10^{-6})} = 5.04\,\Omega$ or about $5\,\Omega$ **(3)**

5 $130\,km\,h^{-1} = 130\,000\,m \div 3600\,s = 36\,m\,s^{-1}$ **(2)**

2. Estimating physical quantities

1 B – $0.532\,\mu m$ **(1)**

2 1 year = 365 × 24 × 3600 seconds = $3.15 \times 10^7\,s$ or about 32 Ms **(2)**

3 Method: Let's assume the town has 20 000 residents in 5000 homes plus shops, factories and schools, etc. There will be periods of high demand, e.g. evenings, and low demand, e.g. at night. During the day, people may be away from home but will still account for a share of what is being used in the town. A reasonable estimate per person would be about a kilowatt averaged over the day, so about 20 000 × 1000 = $2 \times 10^7\,W$ or 20 MW. **(3)**

4 An object falling 2 m will acquire a velocity of $v = \sqrt{2 \times 9.81 \times 2} = 6.3\,m\,s^{-1}$ (using $v^2 - u^2 = 2as$).
If you bend your legs as you land and stop over ~0.5 m, your deceleration is $\frac{v^2}{2s} = 39.2\,m\,s^{-2}$. If your mass is 80 kg, using $F = ma$ gives a decelerating force of 3100 N and in addition your weight is another 780 N, so a total force of a little under 4000 N. **(3)**

5 A train of twelve carriages each with the mass of about ten cars has mass ~200 000 kg (200 tonnes). An express train may travel at $180\,km\,s^{-1}$, or $50\,m\,s^{-1}$. Its KE is $\frac{1}{2}mv^2 = 0.5 \times 200\,000 \times 50^2$ = 250 MJ. So yes, the beam of the LHC has a comparable amount of energy. **(2)**

3. Experimental measurements

1 D – 5.12 ± 0.02 mm **(1)**

2 (a) An error is any deviation of a measurement from its true value. **(1)**

(b) A random error has no pattern, whereas a systematic error is always the same. **(2)**

(c) A random error has an equal chance of being positive or negative, so taking an average will tend to cancel the errors out. However, a systematic error will always bias measurements in the same way so the average too will be biased. **(2)**

(d) The presence of an unexplained non-zero intercept might suggest a systematic error. **(1)**

3 A – 1000 g. The uncertainties are random – i.e. equally likely to be positive or negative – and are therefore likely to cancel each other out on average. **(1)**

4. Combining errors

1 (a) Calibrated means an accurate scale has been established by comparison to a known standard. **(1)**

(b) 5% of 4.7 V = 4.7 × 0.05 = 0.2 V **(2)**

(c) Resolution is the smallest measurement that can be made with an instrument. **(1)**

(d) The reading can only be to the nearest 0.001 A, so when the current being measured is 0.010 A ± 0.001, then the uncertainty due to resolution = $\frac{0.001}{0.010}$ = 0.1 or 10%. However, this does not take into account the accuracy of the calibration of the instrument, so the total uncertainty will be greater. **(3)**

(e) $R = \frac{V}{I} = \frac{4.7}{0.010} = 470\,\Omega$ **(1)**

(f) The combined uncertainty is 5% + 10% = 15%. **(1)**

(g) 15% of $470\,\Omega$ is $71\,\Omega$ **(1)**

2 B – 0.13 ± 0.002 mm. The thickness is $\frac{6.5}{50}$ = 0.13 mm to two significant figures. The uncertainty per sheet is $\frac{0.1}{50}$ = 0.002 mm. **(1)**

5. Graphs

1 (a) 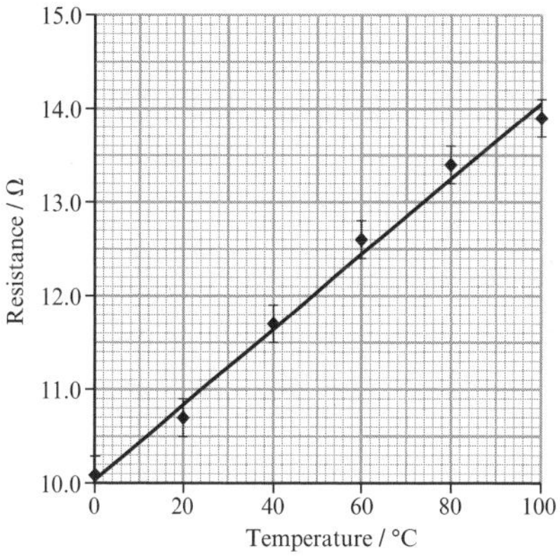

(6)

(b) $0.040\,\Omega^0 C^{-1}$ **(2)**

(c) $10.1 + 0.040 \times 500 = 30\,\Omega$ **(2)**

(d) There are no data for such high temperatures, but the answer assumes that the graph will continue in a linear fashion. **(2)**

6. Scalars and vectors

1

Quantity	Vector	Scalar
Distance		✓
Momentum	✓	
Speed		✓
Energy		✓

(4)

2 (a) Scalar quantities are just magnitudes, so adding them up can only produce one answer, but vectors also have direction. The result of adding two vectors thus depends on their relative directions as well as their magnitudes. **(2)**

(b) 0 N (10 N) (30 N) (70 N) 80 N **(2)**

(c) A – displacement and velocity are vector quantities, but time is a scalar quantity **(1)**

3 Cathy is right, because temperature going up or down is just a change in magnitude; there is no physical direction involved. **(2)**

7. Vector triangles

1 (a)

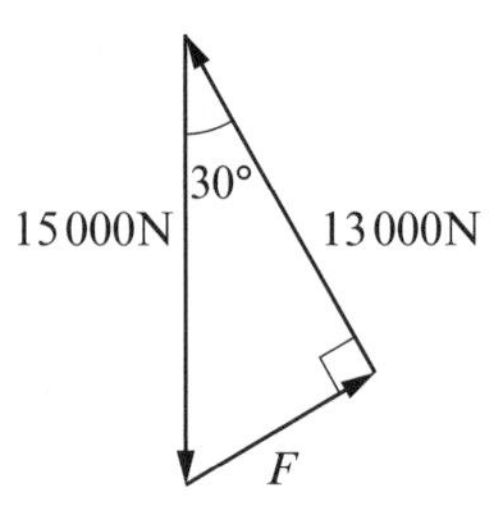

(3)

(b) The arrows all point 'tip to tail' and form a closed triangle. **(2)**

(c) 7500 N **(1)**

2 (a) $\sqrt{3^2 + 5^2} = 5.8$ km **(2)**

(b) 8.0 km @ 4.0 km h^{-1} takes $\frac{8.0}{4.0} = 2.0$ h. 5.8 km @ 3 km h^{-1} takes $\frac{5.8}{3.0} = 1.9$ h, so the saving is 0.1 h or 4 minutes. **(2)**

8. Resolving vectors

1 (a)

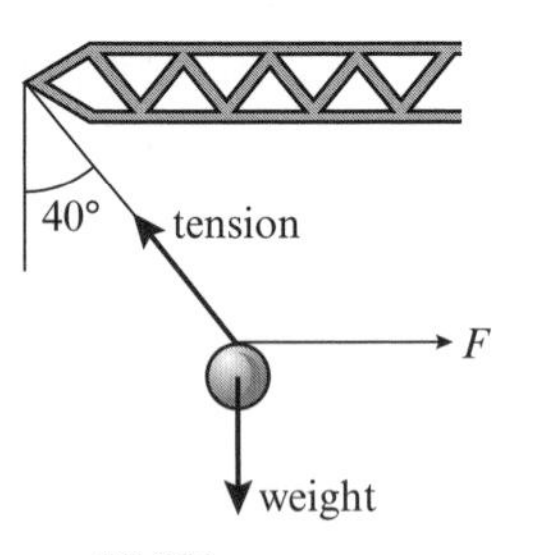

(2)

(b) $T = \frac{20\,000}{\cos 40°} = 26\,100$ N (26.1 kN) **(3)**

(c) $F = 26\,100 \times \cos 50° = 16\,800$ N (16.8 kN) **(2)**

(d) acceleration $a = \frac{(W \cos 50°)}{m_{\text{ball}}} = g \times \cos 50° = 6.3$ m s^{-2} **(2)**

2 B: $F + W \sin \theta$ **(1)**

9. Describing motion

1 (a) Displacement is distance travelled in a particular direction. **(1)**

(b) Velocity is rate of change of displacement. **(1)**

(c) Acceleration is rate of change of velocity. **(1)**

2 (a) average speed $= \frac{120}{6} = 20$ m s^{-1} **(2)**

(b) If an object is accelerating uniformly, its final speed must be twice its average speed, but in this case the final speed = 30 m s^{-1} rather than 20 m s^{-1} **(2)**

(c) average acceleration $= \frac{(30 - 0)}{6} = 5$ m s^{-2} **(2)**

(d) instantaneous speed $= 30 + (40 - 30) \times \frac{4}{8} = 35$ m s^{-1} **(2)**

3 D – the average speed is the mean $= \frac{(u + v)}{2}$ **(1)**

10. Graphs of motion

1 (a)

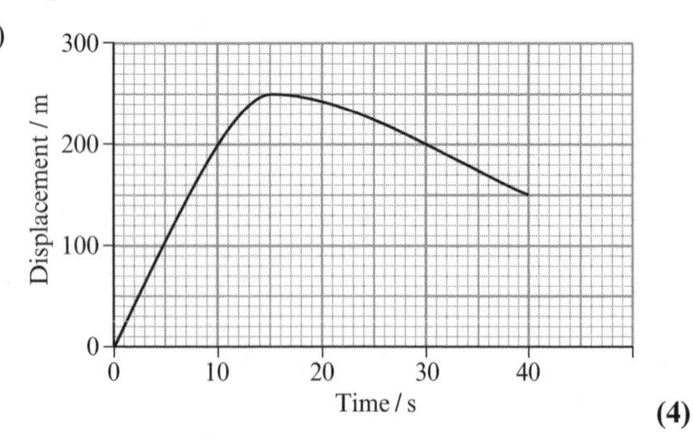

(4)

(b)

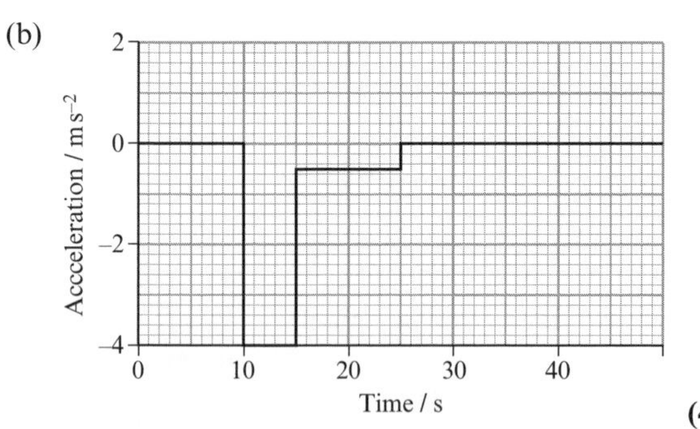

(4)

(c) distance (scalar) = 200 + 50 + 25 + 75 = 350 m **(2)**

11. SUVAT equations of motion

1 (a) If the time is t, B travels $25t$ metres. A (using $s = ut + \frac{1}{2}at^2$) travels $(25t + \frac{1}{2} \times 2 \times t^2)$ metres. A must travel 10 + 15 + 5 + 5 = 35 metres further than B.

$35 + 25t = (25t + \frac{1}{2} \times 2 \times t^2)$

$t^2 = 35$ and so $t = 5.92$ s or just under 6 s. **(3)**

(b) 25 × 5.92 = 148 m (or use $t = 6$ s to get 150 m) **(1)**

(c) 148 + 35 = 183 m (or 150 + 35 = 185 m) **(1)**

(d) $v = u + at = 25 + 2 \times 5.92 = 36.8$ m s^{-1} (or $v = 37$ m s^{-1}) **(2)**

2
- s is a distance so both terms on the right-hand side of the equation must also be distances.
- ut has units of m s^{-1} × s = m and so is a distance.
- The $\frac{1}{2}$ can be ignored (dimensionless).
- at^2 has units of m s^{-2} × s^2 = m and is therefore also a distance term. **(3)**

12. Acceleration of free fall

1 A uniform field has a constant value e.g. 9.81 N kg^{-1} near the surface of the Earth. A heavier object then has a greater downward force acting on it in direct proportion to its mass ($W = mg$); however, acceleration is inversely proportional to mass (Newton's second law). Hence the two effects cancel out and all masses have the same acceleration. **(4)**

2 (a) using $s = ut + \frac{1}{2}at^2$ leads to depth $= 0.5 \times 9.81 \times 2^2 \approx 20$ m **(2)**

(b) The time for sound to travel 20 m is $\frac{20}{340} = 0.06$ s, which is negligibly small compared with 2 s, and there are likely to be much larger timing errors owing to reaction time. **(2)**

3 (a) Using $v^2 - u^2 = 2as$ and setting u as the take-off speed and $v = 0$ at the highest point, $u = \sqrt{2 \times 9.81 \times 0.15} = 1.72$ m s^{-1} **(2)**

(b) $v^2 - u^2 = 2as$ gives $s = \frac{1.72^2}{(2 \times 1.62)} = 0.91$ m **(2)**

(c) $v^2 - u^2 = 2as$ gives $a = \frac{1.72^2}{(2 \times 0.0008)} = 1\,840$ m s^{-2}.

This acceleration is $\frac{1\,850}{9.81} = 189$ times greater than the acceleration of free fall, so the flea would experience 190 times its normal body weight during take-off. This assumes uniform acceleration. **(3)**

13. Vehicle stopping distances

1 C (73 m) – the thinking distance will be twice as long as at the slower speed, as the car travels twice as far in the same time, and the braking distance is proportional to the square of the initial speed ($v^2 - u^2 = 2as$). **(1)**

2 (a) 0.7 s **(1)**

(b) 0.7 × 20 = 14 m **(2)**

(c) 0.5 × 3.0 × 20 = 30 m **(2)**

(d) 14 + 30 = 44 m **(1)**

(e) 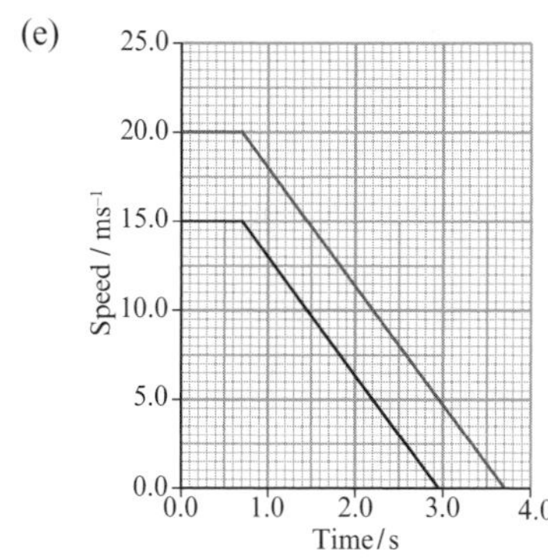

Stopping distance at $15\,\mathrm{m\,s^{-1}} = (15 \times 0.7) + (0.5 \times 15 \times 2.2)$ $= 27.0\,\mathrm{m}$ **(4)**

14. Projectile motion

1 (a) using $s = ut + \frac{1}{2}at^2$ leads to $t = \sqrt{\frac{2s}{a}} = \sqrt{\frac{2 \times 80}{9.81}} = 4.04\,\mathrm{s}$ **(2)**

(b) range = $480 \times 4.04 = 1\,940\,\mathrm{m}$ **(2)**

(c) 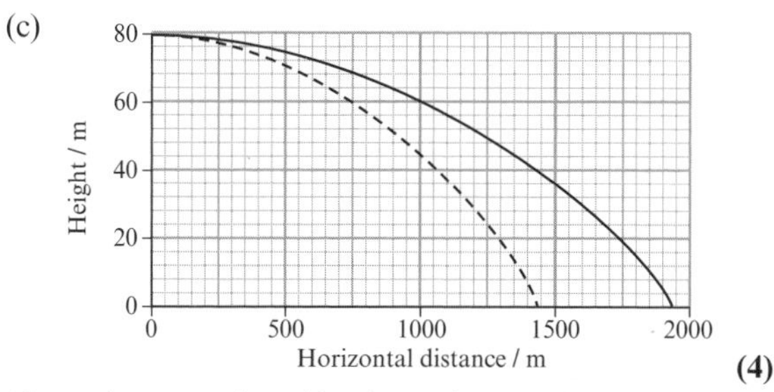

(4)

(d) as above – reduced horizontal range **(2)**

2 Sine has its maximum value of 1 when the angle is 90°. When $\sin 2\theta = 1$, it follows that $2\theta = 90°$ and $\theta = 45°$ for maximum range. **(2)**

15. Types of force

1 (a) C – the force is equal to the weight, so the lift is not accelerating, but you cannot tell whether it is moving upwards or downwards. **(1)**

(b) D – the lift is accelerating downwards, so the upward tension in the newtonmeter spring is reduced. **(1)**

(c) B – $m(g + a)$, because the floor is exerting a contact force greater than the weight, accelerating Isaac upwards. **(1)**

(d) B – $(M + m)(g - a)$, as for (b) above. **(1)**

16. Drag

1 (a) Rearranging the equation above shows that the drag coefficient has the same units as $\frac{\text{force}}{(\text{density} \times \text{area} \times \text{speed}^2)}$ so units are $\frac{\mathrm{kg\,m\,s^{-2}}}{\mathrm{kg\,m^{-3}\,m^2\,m^2\,s^{-2}}}$, which all cancel out to give no units. **(2)**

(b) Drag coefficient is just one factor that affects drag force. In particular, the cross-sectional area will be just as important in determining drag. **(2)**

(c) Because drag force depends on v^2: doubling the speed, for example, will increase the drag by a factor of four. **(2)**

2 (a) Arrow representing drag is shorter:

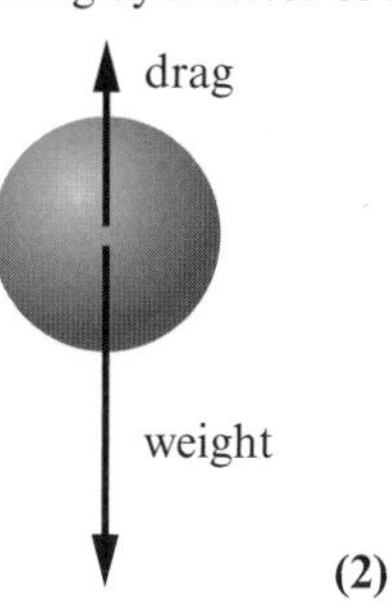

(2)

(b) Both arrows same length:

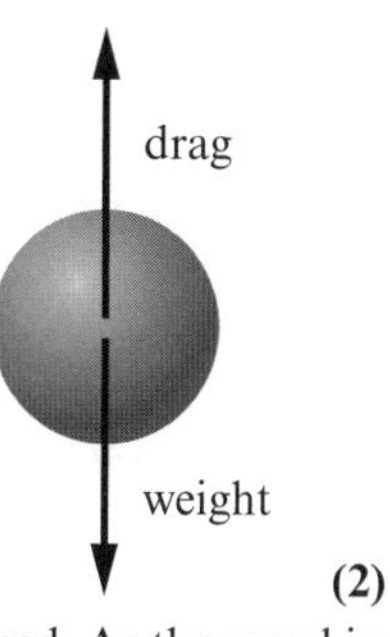

(2)

(c) Drag force depends on speed. As the speed increases, the drag force increases. When drag force = weight, there is zero resultant force and no further acceleration, i.e. terminal velocity has been reached. **(3)**

17. Centre of mass and centre of gravity

1 A **(1)**

2 (a) along centreline in lower half of the cone:

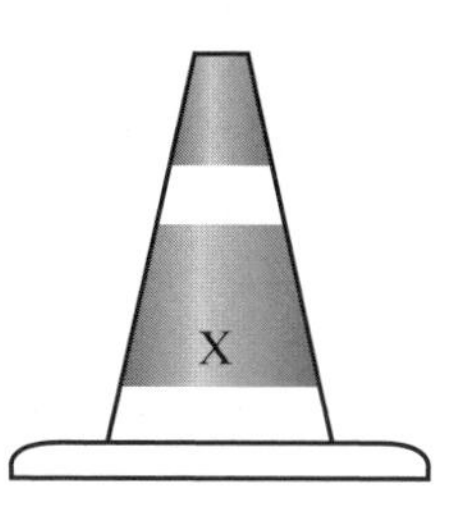

(1)

(b) It has a low centre of mass and a wide base. **(1)**

3 The centre of mass must be vertically below the point of suspension; otherwise a resultant moment will act until the previous condition is satisfied. **(3)**

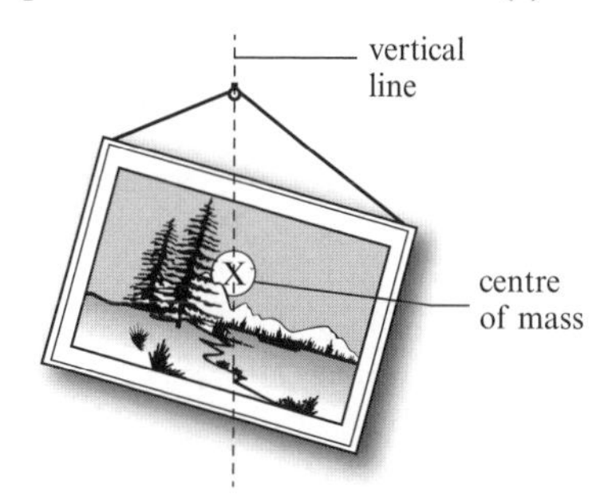

18. Moments, couples and torques

1 (a) Vertically upwards through centre of bolt with an arrow the same length as 50 N arrow:

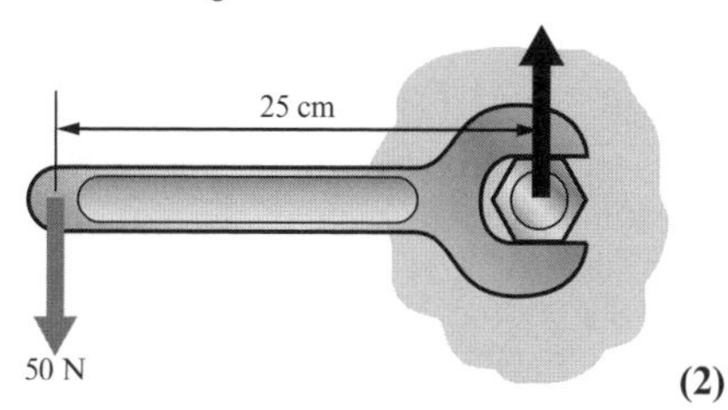

(2)

(b) 50 N **(1)**

(c) $50 \times 0.25 = 12.5\,\mathrm{N\,m}$ **(3)**

2 (a) For an object to be in rotational equilibrium, the total clockwise moment must be equal to the total anticlockwise moment. **(2)**

(b) Considering moments about the wheel of the sack truck, $2.0 \times 0.6 = F \times 1.5$, so $F = 2.0 \times \frac{0.6}{1.5} = 0.8\,\mathrm{kN}$ (800 N). **(2)**

(c) Either principle of moments pivoting about porter's hands: $F \times 0.6 = 0.8 \times 0.9$ giving $F = 1.2\,\mathrm{kN}$
or total downward forces = total upward forces:
$F = 2.0 - 0.8 = 1.2\,\mathrm{kN}$ acting vertically downwards. **(2)**

19. Equilibrium

1 C – his weight is equal in magnitude to the drag force, so there is no resultant force accelerating him **(1)**

2 B

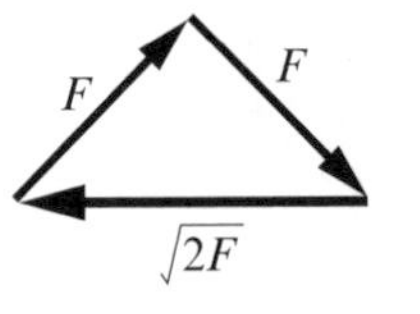

(1)

3 Considering the forces, resultant force = 30 + 20 − 50 = 0 vertically. Considering the torques e.g. about the left-hand end gives resultant torque = (50 × 40) − (20 × 100) = 0. This means there is no resultant force or torque acting on the rule. **(3)**

20. Density and pressure

1 (a) $V = \frac{m}{\rho} = \frac{0.750}{8960} = 8.37 \times 10^{-5}\,\text{m}^3$ **(2)**

(b) The cross-sectional area $= \pi r^2 = \frac{\pi D^2}{4}$

$= \pi \times \frac{(0.80 \times 10^{-3})^2}{4} = 5.03 \times 10^{-7}\,\text{m}^3$ **(2)**

(c) $l = \frac{V}{A} = 8.37 \times \frac{10^{-5}}{(5.03 \times 10^{-7})} = 166\,\text{m}$ **(2)**

2 (a) $p = h\rho g$ leads to $h = \frac{p}{\rho g} = \frac{101\,000}{(13\,600 \times 9.81)} = 0.757\,\text{m}$ **(2)**

(b) The density of water is much lower than that of mercury. A water barometer would have to be 13.6 times taller than a mercury barometer, i.e. over 10 m tall. **(2)**

21. Upthrust and Archimedes' principle

1 (a) Using the equation for hydrostatic pressure: $p = h\rho g$ $= 0.040 \times 1000 \times 9.81 = 392\,\text{Pa}$ **(2)**

(b) $F = pA = 392 \times 0.050^2 = 0.98\,\text{N}$ (alternatively, equate to weight of water displaced) **(2)**

(c) $W = 0.98\,\text{N}$ **(1)**

(d) $m = \frac{W}{g} = 0.98/9.81 = 0.10\,\text{kg}$; $\rho = \frac{m}{V} = \frac{0.10}{0.050^3}$ $= 800\,\text{kg}\,\text{m}^{-3}$ (alternatively, recognise that density of the block must be 80% of that of water) **(2)**

2 (a) $T = mg = 1.00 \times 9.81 = 9.81\,\text{N}$ **(2)**

(b) $T = (1.00 - 0.125) \times 9.81 = 8.58\,\text{N}$ **(2)**

(c) The tension decreases because water is displaced and the mass experiences an upthrust that reduces the tension. It is reduced by the weight of 125 g of water = $0.125 \times 9.81 = 1.23\,\text{N}$ **(2)**

(d) 1500 g **(1)**

22. Exam skills

1 (a) Taking moments about the pivot, the force acting at the support must balance the weight of the barrier acting at its centre of mass $F_{\text{support}} \times 4.0 = 200 \times 2.0$

$F_{\text{support}} = 100\,\text{N}$ **(2)**

(b) The barrier is in force equilibrium, so the downward weight of the barrier must be balanced by the upward force due to the support and the pivot. **(1)**

(c) When the barrier starts to lift, there is zero force between the support and the barrier. If we take moments about the pivot, we get $2.0 \times T\cos 45° = 2.0 \times 200$ and

$T = \frac{200}{\cos 45°} = 280\,\text{N}$ **(3)**

(d) The force at the pivot must balance the horizontal force due to the tension in the wire rope. **(1)**

(e) As the barrier approaches the vertical, the angle between the wire rope and the barrier approaches 90°. This means the component of the tension at right angles to the barrier increases while, at same time, the component of the weight of the barrier at right angles to the barrier decreases. Both effects reduce the tension in the wire rope. **(3)**

(f) Taking moments about the centre of the winch we have $F \times 30.0 = 240 \times 6.0$ so $F = 48\,\text{N}$. **(2)**

23. Work done by a force

1 (a) 400 N **(1)**

(b)

(1)

(c) $W = Fs = 400 \times 5.0 = 2000\,\text{J}$ **(3)**

(d) 2000 J **(1)**

(e) Because the pulling force is not horizontal, only the horizontal component, 460 cos 30°, of the tension in the rope acts against friction, so the total tension must be greater. **(2)**

(f) $W = Fs = (460\cos 30°) \times 5.0 = 2000\,\text{J}$ **(2)**

(g) They are the same because the work done only depends on the frictional force and the distance moved, and both of these are the same in both cases. **(2)**

24. Conservation of energy

1 (a) Energy can't be created or destroyed. It can only be transferred from one form to another. (Alternatively: The total energy of a closed system remains constant.) **(2)**

(b) When energy is transferred, some of it is usually transferred into heat energy. This energy is lost to the surroundings and cannot be reused. This fact means that our supply of useful energy is finite and should not be wasted. **(2)**

2 (a) microphone – sound energy into electrical energy **(2)**

(b) LED – electrical energy into light energy **(2)**

(c) candle – chemical energy into light energy and thermal energy **(2)**

3 (a) 100 − 73 − 8 − 7 = 12% **(1)**

(b) electrical output = 12% of $750\,\text{W}\,\text{m}^{-2} = 0.12 \times 750 =$ $90\,\text{W}\,\text{m}^{-2}$. The area required for 1.8 kW is $\frac{1800}{90} = 20\,\text{m}^2$ **(3)**

(c) advantages: no greenhouse gases, renewable, do not consume useful hydrocarbon feedstock or require transport of fuel; disadvantages: large area needed, dependent on weather, cloud cover, time of day **(4)**

25. Kinetic and gravitational potential energy

1 $v^2 - u^2 = 2as$

If object starts from rest, then $v^2 = 2as$.

The energy of motion is equal to the work done by the accelerating force, W.

$W = Fs$, and $F = ma$, so $W = mas$

as $as = \frac{1}{2}mv^2$, $W = E_k = \frac{1}{2}mv^2$ **(3)**

2 (a) $E_k = \frac{1}{2}mv^2 = 0.5 \times 1800 \times 20^2 = 360\,000\,\text{J}$ **(1)**

(b) The work done by the brakes = initial kinetic energy of the car = Fs leads to $F = \frac{360\,000}{40} = 9000\,\text{N}$ **(2)**

(c) The kinetic energy of the car is transferred by heating to the brakes. **(2)**

3 (a) After each bounce, the maximum velocity is reduced so the ball has less kinetic energy. **(2)**

(b) If the ball has less KE when leaving contact with the surface, it will have less PE at the top of the bounce, and as $E_P = mgh$, the height h must be less. **(2)**

(c) height is area under v–t line from $t = 0.45$ to 0.80 height $= 0.5 \times 0.35 \times 3.4 = 0.60\,\text{m}$ **(2)**

(d) Height of first bounce is 0.60 m which is 60% of the 1 m from which it was dropped, so PE has decreased by 40% and so KE has also decreased by 40%.

Alternatively, velocity decreases from $4.4\,\text{m}\,\text{s}^{-1}$ to $3.4\,\text{m}\,\text{s}^{-1}$ on bouncing, therefore KE decreases to $\left(\frac{3.4}{4.4}\right)^2 = 0.60$ or 60% of initial value. It must therefore have lost 40% of its initial KE. **(2)**

26. Mechanical power and efficiency

1 C cannot be broken down to the base units of $kg\,m^2\,s^{-3}$ **(1)**

2 Work done is equal to force × distance, so the rate at which work is done (the power) is equal to force × $\frac{\text{distance}}{\text{time}}$ = force × velocity. **(2)**

3 (a) (i) $F = \frac{P}{v} = 8 \times \frac{10^3}{20} = 400\,\text{N}$ **(2)**

(ii) $F = \frac{P}{v} = 22 \times \frac{10^3}{30} = 730\,\text{N}$ **(1)**

(b) Drag force increases as the car's speed increases so driving force must also increase to balance it. **(2)**

(c) The time taken to travel 40 km @ 20 km s^{-1}

$= \frac{40\,000}{20} = 2000\,\text{s}.$

The output energy is 8 kW × 2000 s = 16 MJ.

If the efficiency is 20%, the input energy is $\frac{16}{0.20} = 80\,\text{MJ}$.

1 kg of fuel provides 40 MJ so $\frac{80}{40} = 2$ kg of fuel is needed. **(4)**

(d) power = rate of gain of GPE = $mg\frac{\Delta h}{\Delta t}$

$= 1600 \times 9.81 \times 20 \sin 5° = 27\,\text{kW}$ **(3)**

27. Exam skills

1 (a) increase in GPE= $mgh = 2500 \times 9.81 \times 4.0 = 9.8 \times 10^4\,\text{J}$ **(2)**

(b) $9.8 \times 10^4\,\text{J}$ **(1)**

(c) work done by force between pile and ground = $F \times 0.12 = 0.80 \times 9.8 \times 10^4$ so $F = 6.5 \times 10^5\,\text{N}$ **(3)**

(d) thermal energy **(1)**

(e) average power output = $\frac{40}{60} \times 120 \times 10^3 = 8.0 \times 10^4\,\text{W}$ **(2)**

(f) input power = $\frac{10 \times 48 \times 10^6}{3600} = 1.33 \times 10^5$ so $1.3 \times 10^5\,\text{W}$

efficiency = $\frac{8.0 \times 10^4}{1.33 \times 10^5} = 0.60$ or 60% **(2)**

28. Elastic and plastic deformation

1 When elastic deformation occurs, the elastic material returns to its original shape when the deforming force is removed, but when plastic deformation occurs, the deformation is permanent. **(3)**

2 (a) load = 7.0 N, extension = 28 mm **(2)**

(b) the force constant k of a spring is defined by the equation $F = kx$, so $k = \frac{P}{x}$ or the slope of the straight part of the graph. $\frac{P}{x} = \frac{7.0}{0.028} = 250\,\text{N}\,\text{m}^{-1}$ **(2)**

(c) $E_k = \frac{1}{2}kx^2 = 0.5 \times 250 \times 0.020^2 = 0.05\,\text{J}$ **(2)**

3 (a) 22 N **(1)**

(b) $W = Fs = 26 \times 0.080 = 2.1\,\text{J}$ **(2)**

(c) It becomes heat energy that is lost to the surroundings. **(2)**

29. Stretching things

1 At least six points from:
- add a ruler and a set square to the diagram
- clamp ruler and ensure it is vertical using plumb line or spirit level
- use the set square to reduce parallax error
- add masses one by one, noting the extension using the ruler each time
- remove masses one by one to ensure elastic limit was not exceeded
- determine the load for each mass using $W = mg$
- plot a graph of load against extension
- determine the gradient – this is the force constant

(or any additional or equivalent valid points up to a maximum of six marks). **(6)**

2 (a) A long wire will extend more for a given load, reducing the percentage measurement error and so improving accuracy. Similarly, using a thinner wire will increase the extension for a given load. **(3)**

(b) If the wire is very thin, the measurement of its diameter will result in a large percentage error, which is further increased when diameter is squared to calculate the area of cross-section. **(2)**

30. Force–extension graphs

1 (a) The graph is not a straight line through the origin so force and extension are not directly proportional as they are for materials subject to Hooke's law. **(2)**

(b) The unstretching curve passes through the origin, which means the rubber band returns to its original length. **(2)**

(c) It is initially quite stiff up to about 3.0 N then has a lower, fairly constant stiffness until it is loaded with about 5.0 N when it becomes more stiff again. **(2)**

(d) estimated area under the stretching curve leads to work done is ~0.0 52 J **(2)**

(e) The area underneath the stretching curve, which is equal to the work done in stretching the band, is greater than the area under the unstretching curve, which is equal to the work done in unstretching. **(2)**

(f) As a car drives along, the rubber of the tyres is repeatedly deformed and each time it goes through a cycle of stretching and unstretching or compressing and uncompressing, some energy is wasted by heating the tyres. **(2)**

31. Stress and strain

1 (a) C – The tensile stress is inversely proportional to the area, i.e. the square of the diameter. **(1)**

(b) C **(1)**

2 (a) 2000 N **(1)**

(b) equating the vertical forces gives $2T\cos 60° = 2000$ and so $T = \frac{2000}{2\cos 60°} = 2000\,\text{N}$ **(2)**

(c) Stress in all cables is $\frac{2000}{20 \times 10^{-6}} = 10^8\,\text{Pa} = 100\,\text{MPa}$ so no cables would fail. **(3)**

(d) Maximum stress = 50% of 800 MPa = 400 MPa so maximum load is 4 × 2000 N = 8000 N. **(2)**

32. Stress–strain graphs and the Young modulus

1 C **(1)**

2 (a) Tensile stress is the (tensile) force per unit area within the material. It is therefore defined in a similar way to pressure = $\frac{\text{force}}{\text{area}}$ and hence has the same units. **(2)**

(b) Tensile strain is the ratio of extension to original length. It is therefore the ratio of two lengths and has no units. **(2)**

(c) Young modulus is defined as stress/strain, and as strain has no units, the Young modulus and stress must have the same units. **(2)**

3 (a) yield point marked on line at 280 MPa
The yield stress is the stress required to start to deform a material permanently i.e. plastically. **(2)**

(b) Young modulus = $\frac{\text{stress}}{\text{strain}}$ = gradient of the linear part of the graph = $\frac{200 \times 10^6}{0.0030} = 67\,\text{GPa}$ **(2)**

(c) permanent strain = 0.0205 so extension is 100 × 0.0205 = 2.05 mm and new length is 102 mm **(3)**

33. Measuring the Young modulus

1 (a) At least five points from:
- tape measure/metre rule to measure the length of the wire
- micrometer/Vernier or digital callipers to measure the diameter of the wire
- callipers/travelling microscope to measure extension
- repeat each measurement at least three times and take the mean
- fix tape to wire (at pulley end) to act as a reference mark

- use a set-square to reduce parallax error
- add masses one by one and note mass and resultant extension
- remove masses and check that the wire has not moved or stretched permanently

(or any additional or equivalent valid points up to a maximum of five marks). **(5)**

(b) • calculate tensile stress (using $F = mg$ and $\sigma = \frac{F}{A}$)

- calculate strain using ($\varepsilon = \frac{\delta l}{l}$)
- plot a graph of stress against strain
- add a line of best fit
- determine the gradient – this is the Young modulus of copper

(or plot load against extension and Young modulus = gradient $\times \frac{gl}{A}$). **(3)**

2 (a) $\sigma = \frac{10}{\pi \times (0.25 \times 10^{-3})^2} = 51\,\text{MPa}$ **(2)**

(b) $x = \frac{\sigma l}{E} = \frac{3 \times 51 \times 10^6}{120 \times 10^9} = 1.3 \times 10^{-3}\,\text{m} = 1.3\,\text{mm}$

(c) $F = \sigma A = 70 \times 10^6 \times \pi \times (0.25 \times 10^{-3})^2 = 14\,\text{N}$ (or simply $\frac{70}{51} \times 10 = 14\,\text{N}$ **(2)**

(d) Once the wire exceeds the yield point and deforms plastically, it is no longer possible to use the Young modulus to predict extension, as this can only be used when the wire is behaving in a linear elastic manner. **(2)**

34. Exam skills

1 (a) If a single wire were measured and its temperature were to increase, it would increase in length, leading to an error. When two wires are used, both wires expand together, and no temperature-related extension is recorded even though both wires expand. **(2)**

(b)

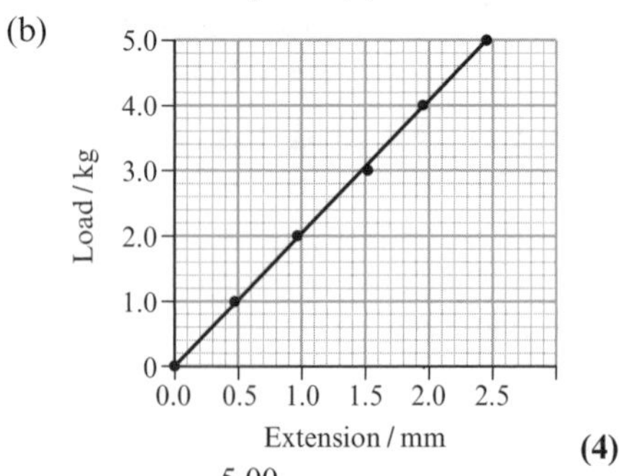

(4)

(c) gradient $= \frac{5.00}{2.46} = 2.03\,\text{kg}\,\text{mm}^{-1}$ **(2)**

(d) radius = 0.30 mm = 0.000 30 m; cross-sectional area = $\pi \times 0.000\,30^2 = 2.8 \times 10^{-7}\,\text{m}^2$ **(2)**

(e) Young modulus $= \frac{\sigma}{\varepsilon} = \frac{F}{A} \times \frac{l}{x} = \frac{mg}{A} \times \frac{l}{x}$. The gradient is $\frac{m}{x}$ so $E = \frac{2.03 \times 10^3 \times 9.81 \times 3.00}{(2.8 \times 10^{-7})} = 2.1 \times 10^{11}\,\text{Pa}$ **(2)**

35. Newton's laws of motion

1 D – Newton's first law says that a body will remain at rest or continue to move at constant velocity until an external force acts on it. **(1)**

2 C – For the ball to change direction there must be a resultant force acting upwards on it. **(1)**

3 C – The lift is accelerating downwards, so the reading on the scales will be less than the weight of the man – there is a net downward force acting on him. **(1)**

4 A – There must be a resultant force acting on the satellite since it is constantly accelerating (changing direction). **(1)**

36. Linear momentum

1 B – check that you have converted all values to base units **(1)**

2 A – Pa s = $\text{kg}\,\text{m}^{-1}\,\text{s}^{-2}$; the others are all $\text{kg}\,\text{m}\,\text{s}^{-1}$ **(1)**

3 B – Iron is accelerated from 0 to $4\,\text{m}\,\text{s}^{-1}$, so over 10 s average a is $0.4\,\text{m}\,\text{s}^{-2}$, so $F = 500 \times 0.4 = 200\,\text{N}$. **(1)**

4 (a) time $= \frac{\text{distance}}{\text{average speed}} = \frac{0.60}{16} = 0.10\,\text{s}$ (average speed while decelerating uniformly = $6\,\text{m}\,\text{s}^{-1}$) **(2)**

(b) momentum = $1500 \times 12 = 18\,000\,\text{kg}\,\text{m}\,\text{s}^{-1}$ **(2)**

(c) force $= \frac{\text{momentum change}}{\text{time}} = \frac{18\,000}{0.10} = 180\,000\,\text{N}$ **(2)**

(d) The mass of the driver is only a fraction of the mass of the car so has a much lower initial momentum. The driver is also likely to be wearing a seatbelt, which will increase the time that he or she takes to slow down and hence reduces the force. **(2)**

37. Impulse

1 (a) $p = mv = 0.600 \times 40 = 24\,\text{kg}\,\text{m}\,\text{s}^{-1}$ **(2)**

(b) impulse is the product of force and the time for which the force is applied. **(1)**

(c) impulse = change in momentum = 24 − 0 = 24 N s **(1)**

(d) force $= \frac{\text{impulse}}{\text{time}} = \frac{24}{0.0005} = 48\,000\,\text{N}$ **(2)**

(e) average speed of hammer as it decelerates = $20\,\text{m}\,\text{s}^{-1}$ and distance travelled by hammer head (= distance travelled by nail) = speed × time = $20 \times 0.0005 = 0.010\,\text{m}$ **(2)**

2 (a) reaction time = 180 ms (0.180 s) **(1)**

(b) 400 ms (0.400 s) **(1)**

(c) peak acceleration $= \frac{\text{peak force}}{\text{mass}} = \frac{1700}{80} = 21.3\,\text{m}\,\text{s}^{-2}$ **(2)**

(d) The area under the force–time graph gives the change in momentum of the sprinter. Each small square is equal to $100\,\text{N} \times 0.020\,\text{s} = 2.00\,\text{N}\,\text{s}$, so the total change in momentum (impulse) is estimated as the area = 380 N s ± 10 N s i.e. $380\,\text{kg}\,\text{m}\,\text{s}^{-1}$. The speed gained is $\frac{380}{80} = 4.8\,\text{m}\,\text{s}^{-1}$. **(3)**

38. Conservation of linear momentum–collisions in one dimension

1 C – P is stationary and Q moves off at $2\,\text{m}\,\text{s}^{-1}$ **(1)**

2 D – $0.5\,\text{m}\,\text{s}^{-1}$ to the left, so total momentum is still 0 **(1)**

3 (a) conservation of momentum gives total momentum before impact = total momentum after impact = $0.200 \times 31 = 0.200 \times v_{club} + 0.045 \times 60$, so $v_{club} = 17.5\,\text{m}\,\text{s}^{-1}$ to the right **(2)**

(b) $F = \left(\frac{\Delta mv}{\Delta t}\right)_{ball} = \frac{0.045 \times 60}{0.0005} = 5400\,\text{N}$ **(2)**

39. Collisions in two dimensions

1 (a) There is no force acting on the puck in a direction parallel to the wall so the component of momentum parallel to the wall must remain constant. **(2)**

(b) The perpendicular component will be $mv \sin\theta = 0.180 \times 20 \times \sin 40° = 2.31\,\text{kg}\,\text{m}\,\text{s}^{-1}$ **(2)**

(c) $2.31\,\text{kg}\,\text{m}\,\text{s}^{-1}$ **(1)**

(d) $\Delta p = -2.31 - (2.31) = -4.62\,\text{kg}\,\text{m}\,\text{s}^{-1}$ **(2)**

2 D – there must be a component of momentum down the page to balance the upward component of Y. **(1)**

40. Elastic and inelastic collisions

1 B – momentum is always conserved, but kinetic energy is conserved only in perfectly elastic collisions. **(1)**

2 (a) velocity = momentum before collision/total mass after the collision $= \frac{1500 \times 10}{(1500 + 2500)} = 3.8\,\text{m}\,\text{s}^{-1}$ ($3.75\,\text{m}\,\text{s}^{-1}$) **(2)**

(b) $E_k = \frac{1}{2}mv^2 = \frac{1}{2} \times 1500 \times 10^2 = 75\,000\,\text{J}$ **(1)**

(c) $E_k = \frac{1}{2}mv^2 = \frac{1}{2} \times 4000 \times 3.75^2 = 28\,000\,\text{J}$ **(1)**

(d) The collision is inelastic as some of the KE is converted into heat and sound energy. **(2)**

3 B – $E_k = \frac{p^2}{2m} = \frac{m^2v^2}{2m}$ **(1)**

41. Electric charge and current

1 C – ampere **(1)**

2 C – The same amount of charge flows through lamp A and lamp B in any given time. **(1)**

3 the number of electrons per second = charge per second (i.e. current)/charge per electron = $\frac{0.004}{(1.60 \times 10^{-19})} = 2.5 \times 10^{16}$ **(2)**

4 (a) $Q = It = 1.8 \times 60 \times 60 = 6480\,C$ per cell, so total $Q = 2 \times 6480 = 13\,000\,C$ **(2)**

(b) $I = \frac{Q}{t} = \frac{(2 \times 6480)}{(4 \times 3600)} = 0.90\,A\ (900\,mA)$ **(1)**

(c) $t = \frac{Q}{I} = \frac{(2 \times 6480)}{0.20} = 64\,800\,s$ (18 hours) **(1)**

42. Charge flow in conductors

1 Conduction requires mobile charge carriers. Metals contain many delocalised electrons that act as charge carriers and can move through the metal. **(2)**

2 The resistance of a metallic conductor increases with increasing temperature. This is because lattice vibration increases with increasing temperature, resulting in conduction electrons losing more energy to the lattice. **(3)**

3 Glass is an insulator at room temperature because the chemical bonding in glass involves all available electrons leading to there being no free charge carriers. **(2)**

4 (a) anode on the left; cathode on the right. **(1)**

(b) D – electrons in the wire **(1)**

(c) C – positive and negative ions **(1)**

(d) A – positive ions **(1)**

43. Kirchoff's first law

1 4 A away from the central node **(2)**

2

LED combination	I_1 / mA	I_2 / mA	I_3 / mA	I_4 / mA
Red	20	20	0	0
Red + amber	45	20	25	0
Green	18	0	18	18
Amber	25	0	25	0

(4)

3 (a) $R_1 = \frac{V}{I_1} = 6.0/0.5 = 12\,\Omega$ **(1)**

(b) current in R_2 is $2.0 - 0.5 = 1.5\,A$ so $R_2 = \frac{6.0}{1.5} = 4.0\,\Omega$ **(2)**

44. Charge carriers and current

1 (a) $10 \times 8.96 = 89.6\,g$ **(1)**

(b) number of moles = $\frac{\text{mass in g}}{\text{molar mass in g mol}^{-1}} : \frac{89.6}{63.5} = 1.41$ mol **(1)**

(c) $1.41 \times 6.02 \times 10^{23} = 8.49 \times 10^{23}$ **(1)**

(d) 8.49×10^{23} **(1)**

(e) $\frac{8.49 \times 10^{23} \times 10^{6}}{10} = 8.49 \times 10^{28}$ **(1)**

(f) $I = Anev$ hence $v = \frac{I}{nAe} = \frac{100}{(8.49 \times 10^{28} \times 10^{-4} \times 1.60 \times 10^{-19})}$ $= 7.36 \times 10^{-5}\,m\,s^{-1}$ **(2)**

45. Electromotive force and potential difference

1 The e.m.f. of a battery is the energy gained per unit charge by charges passing through the battery when chemical energy is transferred to electrical energy. **(2)**

2 D – $kg\,m^2\,s^{-3}\,A^{-1} = J\,C^{-1}$ **(1)**

3 (a) $Q = It = 0.050 \times 100 = 5.0\,C$ **(2)**

(b) $W = VIt = VQ = 9 \times 5.0 = 45\,J$ **(2)**

(c) $6 \times 5.0 = 30\,J$ **(1)**

(d) Energy must be conserved, so if the battery is producing 9 J of electrical energy per coulomb of charge and 6 J is converted to heat and light by the lamp per coulomb, there must be $9 - 6 = 3\,J$ remaining per coulomb and this must be equal to the potential difference across the resistor. **(3)**

46. Resistance and Ohm's law

1 Resistance is defined as the ratio of the potential difference across a conductor to the current through it (or $R = \frac{V}{I}$). **(1)**

2 (a) Variable resistor – used to control the current through the tank and hence the potential difference between the electrodes. **(2)**

(b) Any two from: the depth to which the copper electrodes are immersed / the concentration of the copper sulfate solution / the separation of the copper electrodes. **(2)**

(c)

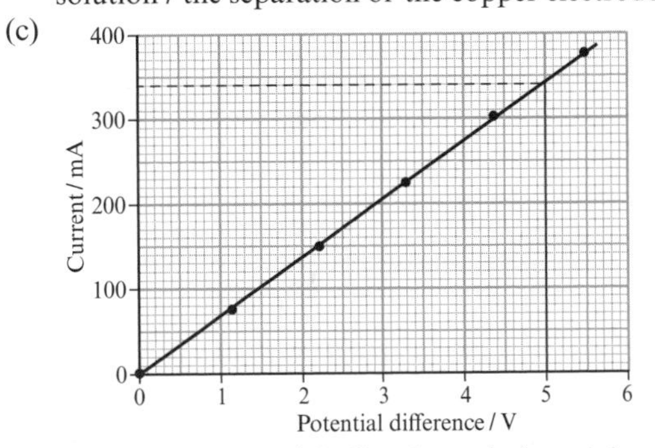

(4)

(d) The graph is a straight line through the origin so it is correct to say the current is directly proportional to potential difference: that is, it obeys Ohm's law. **(2)**

(e) for triangle marked above, $R = \frac{V}{I} = \frac{5.0}{0.34} = 15\,\Omega\ (14.7\,\Omega)$ **(1)**

47. I–V characteristics

1 A and D are ohmic as their $I-V$ characteristics are straight lines that pass through the origin. **(2)**

2 (a) $R = \frac{V}{I} = \frac{2.0}{0.16} = 13\,\Omega$ **(2)**

(b) $R = \frac{V}{I} = \frac{6.0}{0.25} = 24\,\Omega$ **(1)**

(c) The resistance at 2.0 V (0.16 A) is lower than the resistance at 6.0 V (0.25 A) because as the current increases, the temperature of the filament increases, and being a metal, its resistance will also increase. **(2)**

3 No current flows through the diode in the p.d. range from −2.0 V to 0.6 V. Above the threshold value of 0.6 V, the current through the diode increases at an increasing rate until it reaches its maximum value of 400 mA when the p.d. across it is 1.8 V. **(3)**

48. Resistance and resistivity

1 (a) using $R = \frac{V}{I}$ gives $R = \frac{230}{4.35} = 52.9\,\Omega$ **(1)**

(b) $R = \frac{\rho l}{A}$ can be rearranged to give $l = \frac{RA}{\rho} = \frac{52.9 \times 7.55 \times 10^{-8}}{(1.06 \times 10^{-6})} = 3.77\,m$ **(3)**

2 (a) $5.0\,\mu m = 5.0 \times 10^{-6}\,m$ and the twelve wires are in series so using $R = \frac{\rho l}{A}$ gives $\frac{4.9 \times 10^{-7} \times 0.010 \times 12}{(0.10 \times 10^{-3} \times 5.0 \times 10^{-6})} = 120\,\Omega$ **(3)**

(b) The volume is constant, so $V = Al = A \times 0.010 = A' \times 0.01001$ where A' is the area of cross-section after straining. $A' = \frac{A \times 0.010}{0.01001} = 0.999\,A$ so A has decreased by 0.1%. **(2)**

(c) The above result (b) tells us that the wire in the strain gauge gets 0.1% longer and 0.1% thinner in terms of area of cross-section. $R = \frac{\rho l}{A}$ so we know that both of these effects increase the resistance. After straining $R' = \frac{\rho l}{A'}$, so $\frac{R'}{R} = \frac{0.01001}{0.010} \times \frac{1}{0.999} = 1.002$ so R will increase by 0.2%. **(2)**

49. Resistivity and temperature

1 (a) 1.37 **(1)**

(b)

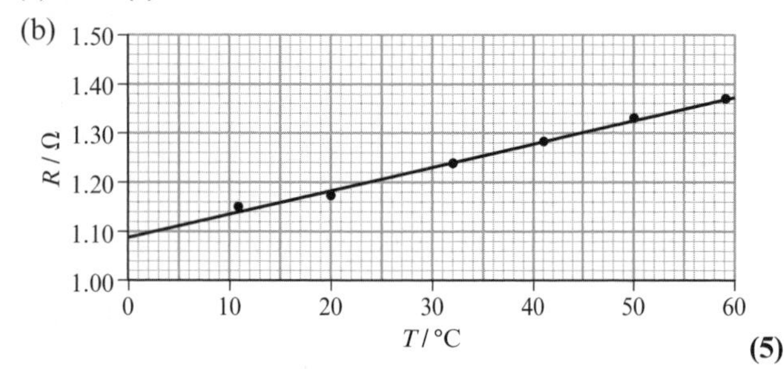

(5)

(c) Gradient is 0.0047 Ω °C^{-1} **(2)**

(d) Line intercepts resistance axis at 1.09 Ω so this is the resistance at 0 °C. **(1)**

(e) The gradient and intercept suggest that R is zero at $\frac{-1.09}{0.0047} = -232$ °C. **(2)**

2 D – negative temperature coefficient means that resistance decreases as temperature rises, but we do not know by how much **(1)**

50. Electrical energy and power

1 (a) The kinetic energy of the electron is equal to the work done by the accelerating voltage, so we can use $E_k = W = VQ = 5000 \times 1.60 \times 10^{-19} = 8.0 \times 10^{-16}$ J. **(2)**

(b) $E_k = \frac{1}{2}mv^2$, so we can find v by rearranging this equation to give $v = \sqrt{\left(\frac{2E_k}{m}\right)}$

so $v = \sqrt{\frac{2 \times 8.0 \times 10^{-16}}{9.11 \times 10^{-31}}} = 4.2 \times 10^7$ m s^{-1} **(2)**

(c) Power is rate of doing work or rate of transfer of energy from one form to another. **(1)**

(d) $P = VI = 5000 \times 2 \times 10^{-3} = 10$ W **(2)**

(e) $P = I^2R = 0.30^2 \times 21 = 1.9$ W **(2)**

2 2000 W = 2 kW and 45 minutes = 0.75 h so in 6 h, the heater uses $6 \times 0.75 \times 2 = 9$ kW h. The total cost is therefore $9 \times 9.5 = 86$ pence. **(3)**

51. Kirchoff's laws and circuit calculations

1 C – Kirchhoff's second law **(1)**

2 (a) D – 1.5 V: there are five cells with an e.m.f. in one direction and one in the opposite direction, so 4 × 1.5 V, and four equal resistors in the circuit. **(1)**

(b) B – 0.30 A (V = 6 V, R = 4 × 5 Ω) **(1)**

(c) D – 0.4 A to the right (consider the left-hand circuit alone: $V = I_1R_1 + I_2R_2$, so $4.5 = 5I_1 + 5 \times 0.5$) **(1)**

(d) A – 0.1 A to the left (solved as for (c), right-hand or whole circuit) **(1)**

52. Resistors in series and parallel

1 if each resistor has resistance R, the total resistances are R, $\frac{2R}{3}$, $\frac{3R}{2}$ and $\frac{6R}{7}$ respectively so:

(a) C has the highest resistance **(1)**

(b) B has the lowest resistance. **(1)**

2 (a) **(1)** (b) **(1)**

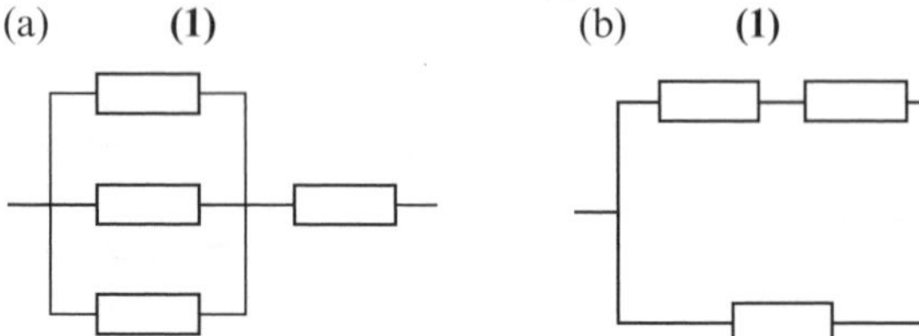

3 (a) The network is equivalent to 10 + 15 + 20 = 45 Ω in parallel with 25 + 30 = 55 Ω which has a total resistance of $\frac{(45 \times 55)}{(45 + 55)} = 24.75$ Ω or 25 Ω to 2 significant figures. **(1)**

(b) The network is equivalent to 10 + 20, 30 and 5 + 25 in parallel and as this is three lots of 30 Ω in parallel, the result is $\frac{30}{3} = 10$ Ω. **(1)**

53. DC circuit analysis

1 (a) from the graph, the p.d. across the LED is 1.7 V **(1)**

(b) The p.d. across R is 3.0 − 1.7 = 1.3 V. Using $R = \frac{V}{I}$ gives $R = \frac{1.3}{0.020} = 65$ Ω. **(2)**

(c) If the p.d. across the diode is 1.1 V, from the graph, the current through it is 6 mA. The p.d. across the 65 Ω resistor is $V = IR = 0.006 \times 65 = 0.39$ V. The e.m.f. of the battery must now be equal to 1.1 + 0.39 = 1.5 V. **(2)**

(d) The p.d. across the first 100 Ω resistor is $V = IR = 0.045 \times 100 = 4.5$ V. The current then splits three ways, so 15 mA flows through each of the other 100 Ω resistors. The p.d. across each of them is 0.015 × 100 = 1.5 V. The current through each LED is also 15 mA and from the characteristic, the p.d. across each LED is also 1.5 V. Considering any one loop that includes the battery means that its e.m.f. must be 4.5 + 1.5 + 1.5 + 1.5 = 9.0 V. **(3)**

54. E.m.f. and internal resistance

1 The current in the circuit is the e.m.f. of the battery divided by the total circuit resistance, including the internal resistance, and it is also equal to the terminal voltage divided by the external resistance. If the current around the circuit is I, then for R, $I = \frac{V}{R}$ and for the whole circuit, $I = \frac{\varepsilon}{R + r}$ so we have $\frac{V}{R} = \frac{\varepsilon}{R + r}$ and therefore $V = \varepsilon \frac{R}{R + r}$. **(2)**

2 B (as the currents are A: $\frac{1.5}{(0.1 + 0.5)} = 2.5$ A, B: $\frac{3.0}{(0.2 + 0.5)} = 4.29$ A, C: $\frac{3.0}{(0.4 + 0.5)} = 3.33$ A and D: $\frac{4.5}{(0.6 + 0.5)} = 4.09$ A respectively). **(1)**

3 (a) $V_r = Ir = 1.2 \times 0.50 = 0.60$ V **(1)**

(b) $V_{lamp} = \varepsilon - V_r = 6.0 - 0.60 = 5.4$ V **(1)**

(c) $P = VI = 5.4 \times 1.2 = 6.5$ W **(1)**

(d) When there is internal resistance (and a current flowing), some p.d. is always 'dropped' across the internal resistance, which means the full battery e.m.f. is not available and the maximum power and hence maximum brightness associated with the bulb's 6 V normal rating cannot be achieved. **(2)**

55. Experimental determination of internal resistance

1 (a) The uncertainty in V is ±0.01 V and the uncertainty in I is ±0.01 A (because digital measurements proceed in steps, whereas with analogue instruments you round to the nearest scale division, so the measurement uncertainty is half the smallest division). **(1)**

(b)

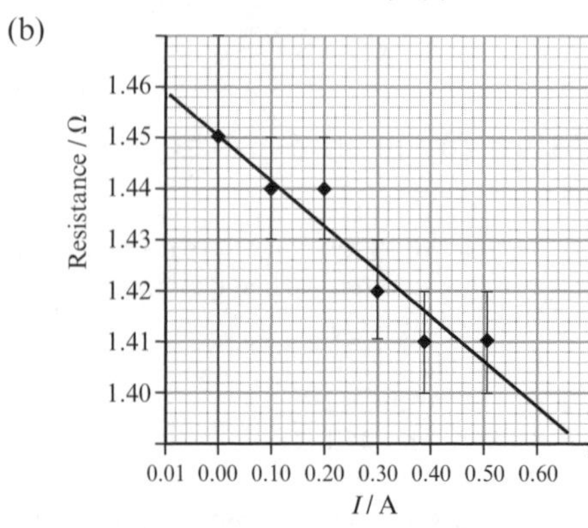

(5)

(c) Given that $\varepsilon = V + Ir$ it follows that $V = -Ir + \varepsilon$ and if we compare this to the equation of a straight line, $y = mx + c$

it follows that a graph of V against I will be a straight line of gradient $-r$. **(2)**

(d) gradient $= \frac{-0.05}{0.57} = -0.088$, hence $r = 0.09\,\Omega$. **(2)**

(e) The e.m.f. of the cell is the y intercept of the graph, so $\varepsilon = 1.45\,V$. **(1)**

(f) The biggest problem is that the voltmeter chosen is not precise enough. It should measure to ±0.001 V in order to eliminate the 'steps'. Six points is also not really enough for a good graph, so the range of measurements should be increased. **(2)**

56. Potential dividers

1 (a) the potential divider formula $V_{out} = V_{in}\frac{R_2}{R_1 + R_2}$ gives $V = \frac{9 \times 30}{(20 + 30)} = 5.4\,V$ **(2)**

(b) the p.d. across the 20 Ω resistor below Y is 9.0 − 5.4 = 3.6 V, so the p.d. between X and Y is 5.4 − 3.6 = 1.8 V **(2)**

2 (a) the p.d. across the 10 kΩ resistor is $\frac{6.0 \times 10}{(10 + 20)} = 2.0\,V$ **(1)**

(b) The voltmeter is in parallel with the 10 kΩ resistor, so the effect is to produce a potential divider consisting of a 20 kΩ resistor in series with 5 kΩ resistor and the voltmeter will read $\frac{6.0 \times 5}{(5+20)} = 1.2\,V$. **(2)**

57. Investigating potential divider circuits

1 (a) The sensitivity decreases as the temperature increases from 0 to 60 °C **(1)**

(b) The sensitivity of the thermistor at 30 °C is equal to the slope of the curve at 30 °C, so a tangent to the curve is needed: at 30 °C a tangent has a gradient of $\frac{3.15}{47}$ $= 0.067\,k\Omega\,°C^{-1}$ ($67\,\Omega\,°C^{-1}$).

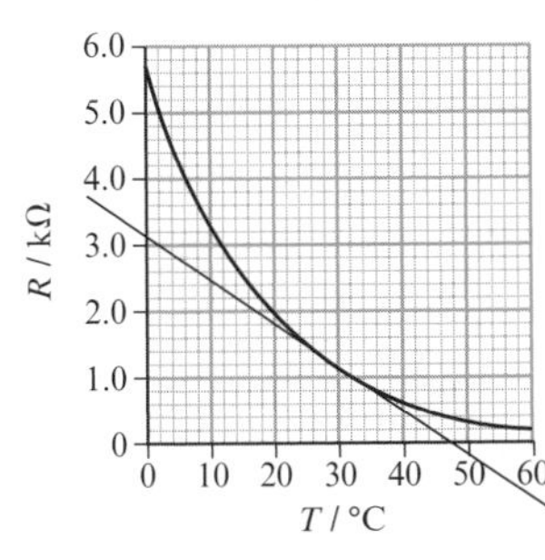

(3)

(c) At 12 °C, the resistance of the thermistor from the graph is 3.0 kΩ, so the output p.d. will be 3.0 V, as both resistances in the potential divider are the same. **(2)**

(d) At 40 °C, from the graph, the resistance of the thermistor is 0.6 kΩ so the output p.d. will be $V = \frac{6.0 \times 3.0}{(3.0 + 0.6)} = 5.0\,V$. **(2)**

(e) $V_{out} = \frac{(6.0 \times 3.0)}{(3.0 + R_t)}$, so $R_t = 3.6 = \frac{6 + 3}{3 + R_t}$, $3 + R_t = \frac{6 \times 3}{3.6} = 5$

$= 2.0\,k\Omega$

$R_t = 5 - 3 = 2.0\,k\Omega$

$R_t = 2\,k\Omega$ at 20 °C. **(2)**

58. Exam skills

1 (a) Electromotive force is the energy transferred by a power supply into electrical form as electrical charge passes through it. **(2)**

(b) maximum current is $I = \frac{\varepsilon}{r} = \frac{12.0}{1.0} = 12\,A$ **(1)**

(c) terminal p.d. = 10.0 V **(1)**

(d) When current flows, there will be a potential difference across the internal resistance or 'lost volts', which must be subtracted from the e.m.f. in order to determine the potential difference across the terminals of the battery. In this case, 2.0 V is dropped across the internal resistance of the battery. **(2)**

(e) p.d. across circuit is 10 V, current $= \frac{\text{lost volts}}{\text{internal resistance}}$ $= \frac{2.0}{1.0} = 2.0\,A$. **(2)**

(f) When the second lamp is added in parallel to the first, more current flows so the size of the lost volts increases and the terminal p.d. of the battery decreases. This means that the p.d. across each lamp will now be less than 10.0 V and they will not light with normal brightness. **(3)**

59. Properties of progressive waves

1 (a) (b)

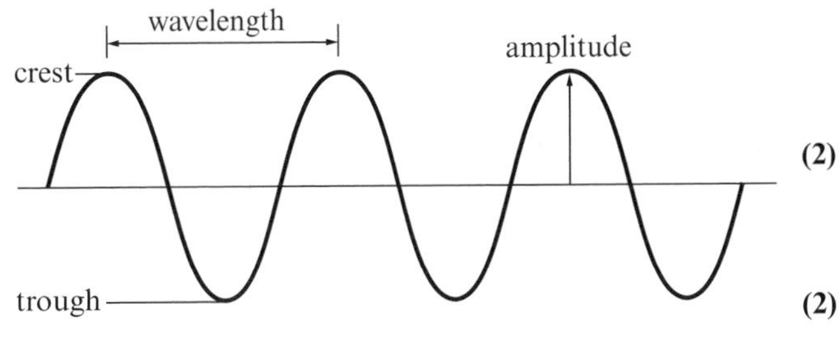

(2)

(2)

2

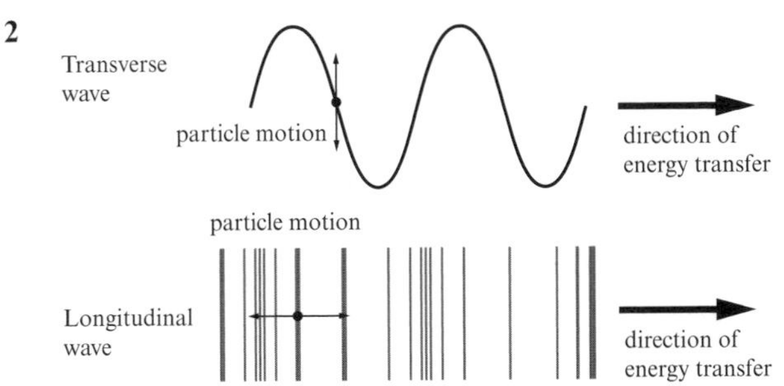

In the case of transverse waves, the particles carrying the wave oscillate at right angles to the direction of propagation of the wave or the direction of energy transfer. For longitudinal waves, the particles oscillate parallel to the direction that the wave is travelling. (In the case of electromagnetic waves that are also transverse, the wave consists of oscillating electric and magnetic fields which are perpendicular to the direction of energy transfer.) **(4)**

3 (a)

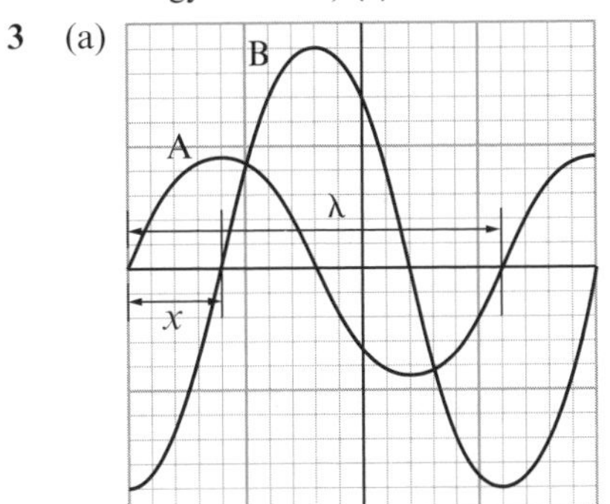

Phase difference $= 2\pi \times \frac{x}{\lambda} = 2\pi \times \frac{4}{16} = \frac{\pi}{2}$ or 90° **(2)**

(b) intensity is proportional to (amplitude)^2 so $\frac{I_B}{I_A} = \frac{9.0^2}{4.5^2} = 4.0$ **(2)**

60. The wave equation

1 Rearranging $v = f\lambda$ gives $\lambda = \frac{v}{f} = \frac{3.00 \times 10^8}{(96.1 \times 10^6)} = 3.12\,m$, so the length should be $\frac{3.12}{4} = 0.78\,m$. **(2)**

2 (a) $f = \frac{v}{\lambda} = \frac{1540}{0.00044} = 3.5 \times 10^6\,Hz$ (3.5 MHz) **(2)**

(b) A very short wavelength will reduce diffraction effects and increase the sharpness and resolution of the ultrasound image. **(2)**

(c) Ultrasound is non-ionising and is essentially harmless to the patient, unlike X-rays. **(1)**

3 (a) $f = \frac{v}{\lambda} = \frac{3.00 \times 10^8}{(532 \times 10^{-9})} = 5.64 \times 10^{14}\,\text{Hz}$ **(2)**

(b)

	In air	In water
Speed / m s^{-1}	3.00×10^8	2.26×10^8
Wavelength / nm	532	400
Frequency / Hz	5.64×10^{14}	5.64×10^{14}

(3)

61. Graphical representation of waves

1 (a) wavelength $\lambda = 8.6\,\text{mm} = 8.6 \times 10^{-3}\,\text{m}$ **(1)**

(b) time period $T = 25\,\mu\text{s} = 2.5 \times 10^{-5}\,\text{s}$, frequency $f = \frac{1}{T}$ $= \frac{1}{(2.5 \times 10^{-5})} = 4.0 \times 10^4\,\text{Hz}$ **(2)**

(c) speed $v = f\lambda = 4.0 \times 10^4 \times 8.6 \times 10^{-3} = 340\,\text{m s}^{-1}$ **(1)**

2

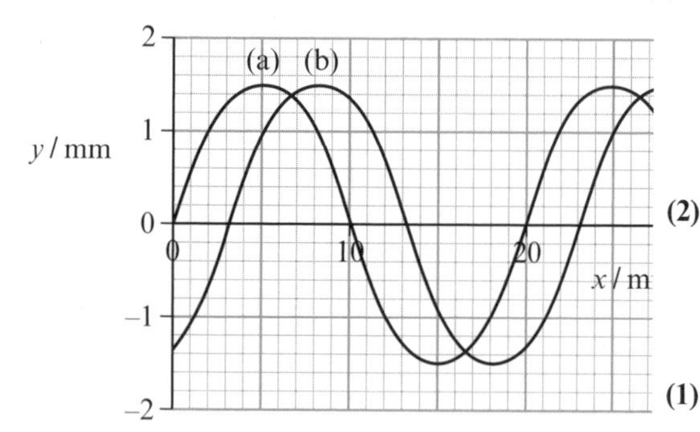

(2)

(1)

62. Using an oscilloscope to display sound waves

1 (a) microphone **(1)**

(b) amplitude = 3.3 divs @ 1 V/ div = 3.3 V **(2)**

(c) one complete wave occupies four divisions or 4 ms, which is equal to 0.004 s, the time period; $f = \frac{1}{t} = \frac{1}{0.004} = 250\,\text{Hz}$ **(2)**

(d)

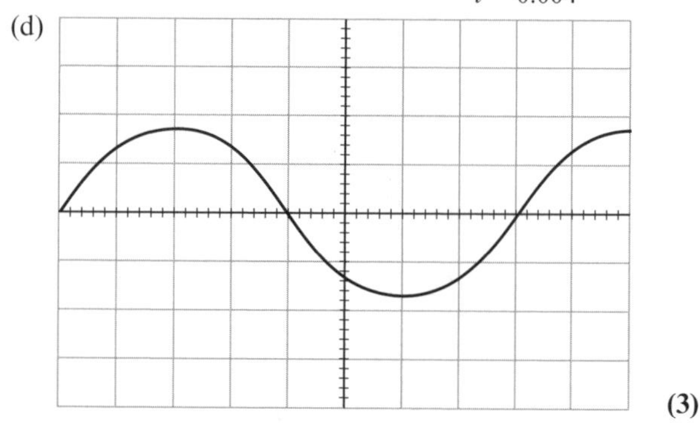

(3)

63 Reflection, refraction and diffraction

1 (a) the speed is reduced but the frequency is unchanged **(1)**
(b) the wavelength is reduced **(1)**
(c) **(3)** (d) **(2)**

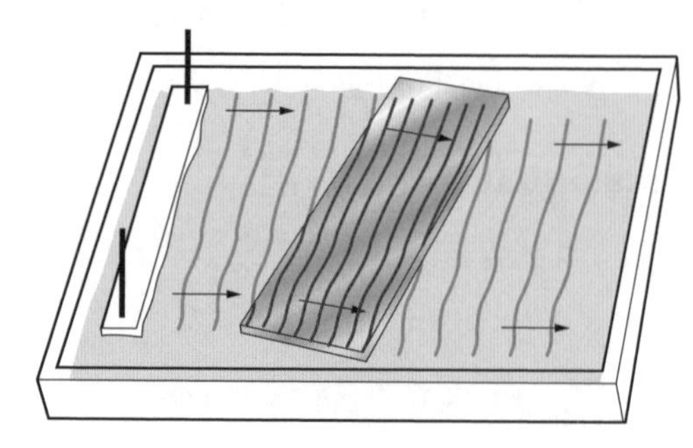

(e)

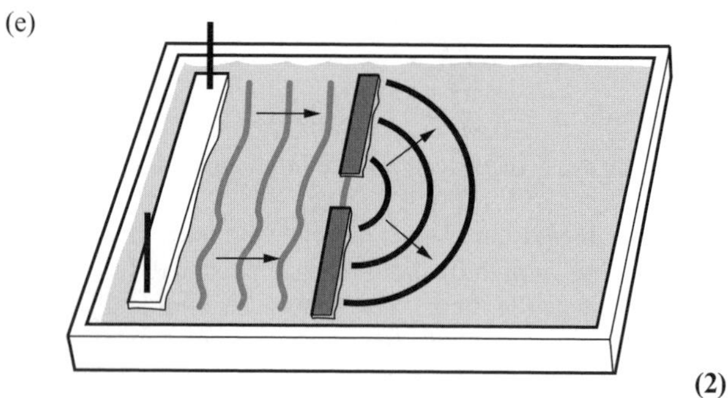

(2)

(f) diffraction **(1)**

64. The electromagnetic spectrum

1 B – ultraviolet (remember visible light has $\lambda = 400–700\,\text{nm}$) **(1)**

2 B – 10 cm (domestic microwave ovens use λ about 12 cm) **(1)**

3 D – longitudinal waves cannot be polarised **(1)**

4 C – ultraviolet has the shortest wavelength and the greatest energy **(1)**

5 The area of a sphere of radius r is given by $A = 4\pi r^2$ and intensity is power/area as the light spreads out over the surface of a sphere. The light output is 7.5% of 100 W, which is 7.5 W,

so the intensity is $I = \frac{P}{A} = \frac{7.5}{(4 \times \pi \times 2.0^2)} = = 0.15\,\text{W m}^{-2}$. **(3)**

6 (a) $\lambda = \frac{v}{f} = \frac{3.00 \times 10^8}{2.45 \times 10^9} = 0.122\,\text{m}$. $3 \times \lambda = 0.366$ m **(2)**

(b) A standing wave involves nodes – points of zero amplitude – and antinodes – points of maximum amplitude. If the food to be heated were not rotated, some parts would receive almost no microwave radiation and remain cold, while others would receive too much and overcook. **(2)**

65. Polarisation

1 C – ultrasound is a longitudinal wave **(1)**

2 (a) The direction of polarisation of light is actually determined by a vector quantity. Unpolarised light oscillates in every direction but can be resolved into two perpendicular components, e.g. vertical and horizontal. 50% will be vertically polarised and 50% will be horizontally polarised. A vertical polarising filter will only let through the vertically polarised component i.e. 50% of the light. **(2)**

(b) Light passing through the first filter will be plane-polarised in a direction at 45° to the second filter. This light can be resolved into components both parallel to and perpendicular to the second filter and the parallel component can be transmitted. **(2)**

(c) Light passing through the first filter will be plane-polarised in a direction perpendicular to the second filter and thus has zero component parallel to the second filter so no light can be transmitted. **(2)**

3

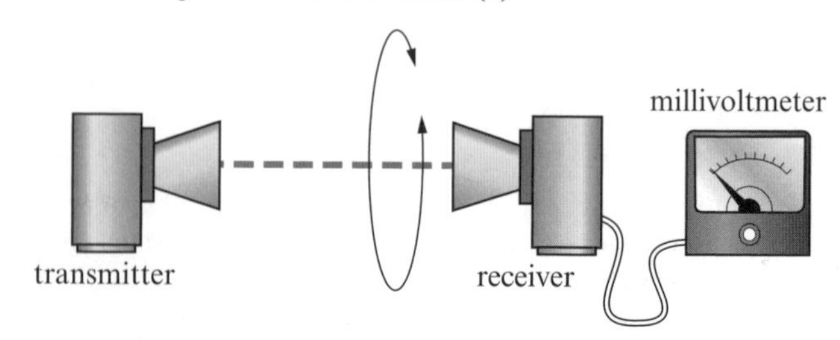

Set up the microwave transmitter and receiver as shown above. A large signal can be detected. The receiver (or transmitter) is now rotated slowly about a horizontal axis. The received signal drops and reaches zero after the receiver has rotated through 90°. Further rotation will restore the signal until it reaches maximum again after rotating 180°. This shows that the microwaves are polarised. **(4)**

66. Refraction and total internal reflection of light

1 $n = \frac{c}{v}$ so $v = \frac{c}{n} = \frac{3.00 \times 10^8}{1.45} = 2.07 \times 10^8\,\mathrm{m\,s^{-1}}$ **(1)**

2 For TIR to occur, the angle of incidence, which is 45°, must be greater than the critical angle. As $\sin C = \frac{1}{n}$, $C = \sin^{-1}\left(\frac{1}{1.49}\right) = 42°$, which is less than 45°, so TIR can occur. **(2)**

3

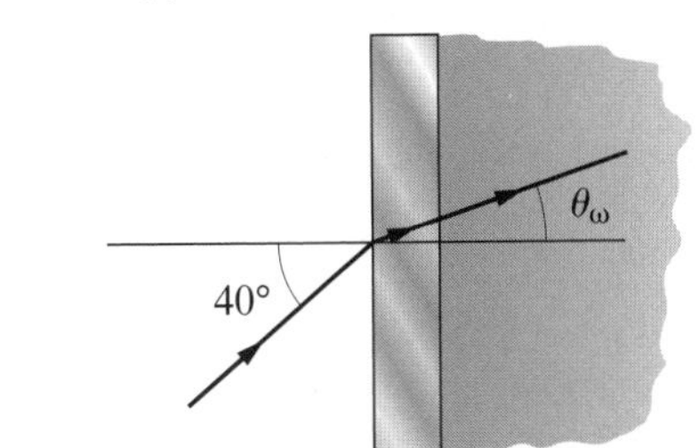

(a) As $n \sin\theta$ = constant, at the air–glass boundary we can say that $n_{air} \sin 40° = n_{glass} \sin\theta_g$ so $\sin\theta_g = \frac{n_{air}\sin 40°}{n_{glass}} = \frac{1.00 \times 0.643}{1.52} = 0.423$ and $\theta_g = 25°$. **(2)**

(b) At the glass–water boundary $n_{glass} \sin\theta_g = n_{water} \sin\theta_w$ so $\sin\theta_w = \frac{n_{glass}\sin\theta_g}{n_{water}} = \frac{1.52 \times 0.423}{1.33} = 0.483$ and $\theta_w = 29°$. **(2)**

(c) $n_{glass} \sin\theta_g = n_{water} \sin\theta_w$ At the critical angle $C = \theta_g$ and $\theta_w = 90°$ and we have $C = \sin^{-1}\left(\frac{1.33}{1.52}\right) = 61°$. **(2)**

(d) For a ray travelling across the air–glass boundary, the biggest angle of refraction is when the angle of incidence is approaching 90° i.e. $\theta_g = \sin^{-1}\left(\frac{1}{1.52}\right) = 41°$, so it could not possibly reach the glass–water boundary at an angle of 61° in order for TIR to occur. **(3)**

67. The principle of superposition

1 C – amplitudes must sum to zero **(1)**
2 C – 2^2 = four times the intensity **(1)**
3 A – the same frequency, f **(1)**
4 D – there is no phase difference over which the two waves can sum to zero **(1)**

5

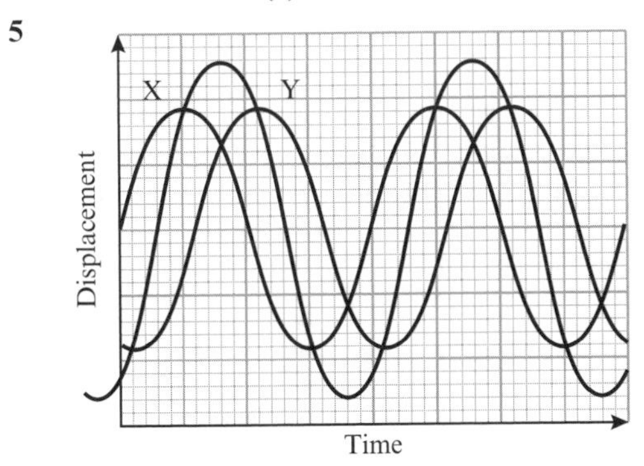

(3)

68. Interference

1 (a) ACB = $\sqrt{70.0^2 + 32.1^2} = 154.0\,\mathrm{cm}$ **(2)**

(b) path difference = 154.0 − 140.0 = 14.0 cm **(1)**

(c) A path difference of one wavelength results in a phase difference of 2π. 14.0 cm = $\frac{14.0}{2.8}$ = 5 wavelengths so the phase difference is $5 \times 2\pi = 10\pi$ radians. **(2)**

(d) The path difference is a whole number of wavelengths so constructive interference occurs at B. **(1)**

2 Firstly, the lamps are not coherent sources, so light will arrive at the screen from the two sources with no predictable phase difference as is needed to create an interference pattern. Secondly, the wavelength of light is so small compared with the spacing of the lamps that the fringe spacing would be extremely small. **(4)**

69. Two-source interference and the nature of light

1 Huygens considered that light propagated as a (longitudinal) wave and required some invisible medium in which to travel (the aether). On the other hand, Newton believed that light consisted of tiny particles or 'corpuscles' moving through a medium. Our current understanding is based on quantum physics and the idea of wave–particle duality. This means that light sometimes exhibits particle-like properties and sometimes wave-like properties but is actually neither. The idea of the aether has now been dismissed as unnecessary and fictitious. **(5)**

2

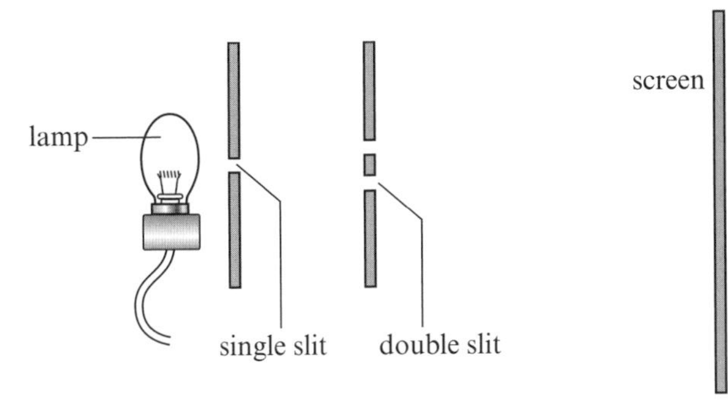

The apparatus is set up as in the diagram with a monochromatic light source behind a single slit. The single slit will ensure that the light travelling towards the double slits is coherent. Light from the double slits superposes and produces a pattern of interference fringes on the screen. Diffraction and interference can only be explained using a wave model, so Young's experiment did precisely that and confirmed that light travels as a wave. **(4)**

3 (a) Two waves are coherent if they maintain a constant phase relationship between each other. **(2)**

(b) Monochromatic sources consist of light of a single colour i.e. frequency which is a necessary condition for the sources to be coherent. The relative phases of two different frequency sources would be constantly changing and so the light would not be coherent. **(2)**

70. Experimental determination of the wavelength of light

1 (a) Ten fringe spacings = 98 mm, so the fringe spacing is 9.8 mm or 9.8×10^{-3} mm. **(2)**

(b) Ten fringe spacings can be measured to ±1 mm (relating to the 'fuzziness' of the fringes) so the fringe spacing iVs to ±0.1 mm. **(2)**

(c) using $\lambda = \frac{ax}{D}$ we have $\lambda = \frac{0.2 \times 10^{-3} \times 9.8 \times 10^{-3}}{3.00}$ $= 6.5 \times 10^{-7}$ m or 650 nm **(3)**

(d) using $n\lambda = d\sin\theta$ we have $\lambda = \frac{\sin 23}{(600 \times 10^3)} = 6.5 \times 10^{-7}$ m; answer to (c) is confirmed. **(2)**

71. Stationary waves

1 Standing waves are created by the superposition of two coherent waves of the same type and of similar amplitude travelling in opposite directions. Furthermore, the waves must have the same wavelength and frequency. **(4)**

2 (a) Microwaves incident on the metal sheet are reflected, creating two waves of similar amplitude but travelling in opposite directions. Where these waves are in phase, constructive interference results in the formation of antinodes, i.e. points of maximum amplitude, and where they are in antiphase, destructive interference results in nodes, i.e. points of zero or minimum amplitude. **(4)**

(b) When the probe encounters an antinode, a maximum output signal is produced because the wave amplitude is a maximum, whereas a node will produce a minimum output signal where the wave amplitude is a minimum. In addition, the maxima increase in amplitude as the probe moves towards the transmitter and the minima become non-zero. This is because the reflected wave gets weaker and destructive interference is not complete as the two amplitudes are no longer the same. **(3)**

(c) The distance between antinodes is half a wavelength so the wavelength is 28 mm. **(2)**

72. Stationary waves on a string

1 (a) C – 4 nodes and 3 antinodes (1.5 wavelengths) **(1)**

(b) C – $\frac{5f}{4}$ as it is not a multiple of $\frac{f}{2}$ **(1)**

2 (a) The string carries two complete waves, so the wavelength is half the length of the string, $\lambda = 0.75$ m, and $f = 16.0$ Hz, so using $v = f\lambda$ gives $v = 16.0 \times 0.75 = 12.0\,\text{m s}^{-1}$. **(2)**

(b) $f = 16.0$ Hz. As the tension increases, the number of antinodes decreases by one, so $\lambda = \frac{2}{3} \times 1.50 = 1.00$ m, so using $v = f\lambda$ gives $v = 16.0 \times 1.00 = 16.0\,\text{m s}^{-1}$. **(2)**

3 D – P and T are always moving in the same direction at the same time **(1)**

73. Stationary sound waves

1 (a) The fundamental mode is the lowest frequency at which resonance can occur: in this case with a node at the bottom and an antinode at the top. **(2)**

(b) The tube length is quarter of a wavelength so $l = 4\lambda$. **(1)**

(c) The missing values are 0.00526 and 0.00588. **(1)**

(d)

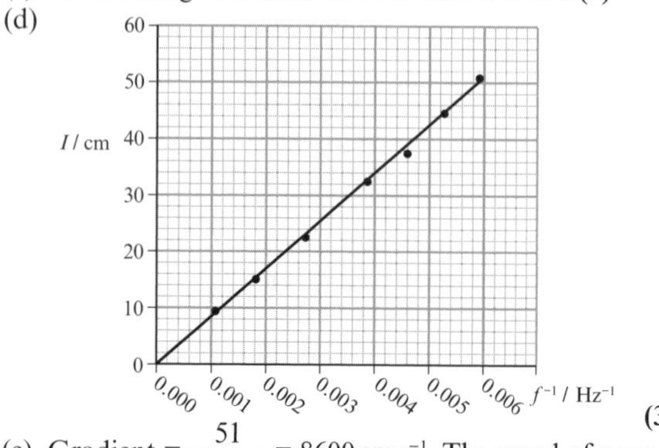

(3)

(e) Gradient $= \frac{51}{0.0060} = 8600\,\text{cm s}^{-1}$. The speed of sound is four times the gradient (because $l = 4\lambda$) and is equal to $34\,000\,\text{cm s}^{-1}$ or $340\,\text{m s}^{-1}$. **(3)**

74. Stationary waves in closed and open tubes

1 (a) **(2)** (b) **(2)** (c) **(1)**

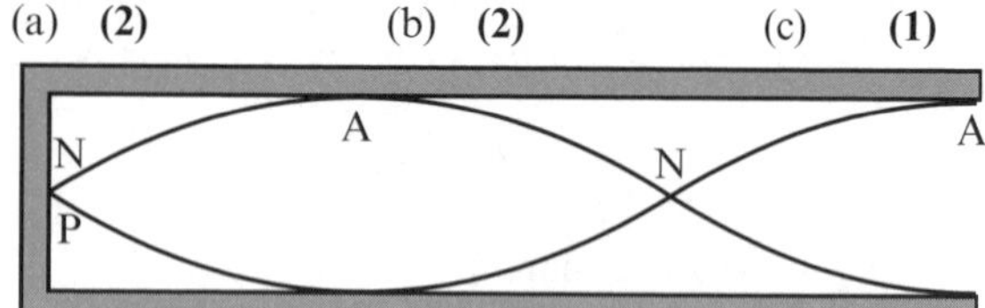

(d) third harmonic, so the length of the pipe is $\frac{3}{4}$ of the wavelength and $\lambda = \frac{4l}{3}$ which means that $f = \frac{v}{\lambda} = \frac{3v}{4l} = 3 \times \frac{340}{(4 \times 0.60)} = 425$ Hz **(2)**

2 C – 72 Hz **(1)**

3 D – 160 Hz **(1)**

4 C – 187 Hz **(1)**

75. Exam skills

1 (a) $n = \frac{\sin i}{\sin r} = \sin^{-1}\left(\frac{\sin 45°}{1.5}\right) = 28°$ **(2)**

(b) The critical angle is the angle of incidence at which the angle of refraction is 90° and the refracted ray travels along the boundary between the two materials. Above this critical angle light does not leave the first material but is entirely reflected at the boundary. **(3)**

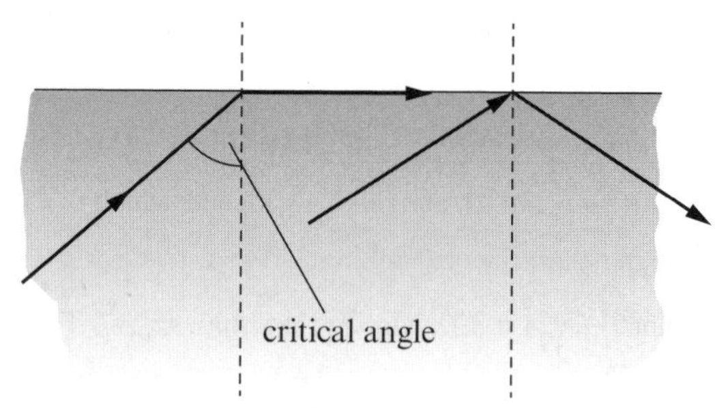

(c) $n = \frac{1}{\sin C}$ so $C = \sin^{-1}\left(\frac{1}{1.5}\right) = 42°$ **(2)**

(d) The angle of incidence at the sloping surface is 28° + 31° = 59°, which is greater than the critical angle so total internal reflection occurs. **(2)**

(e)

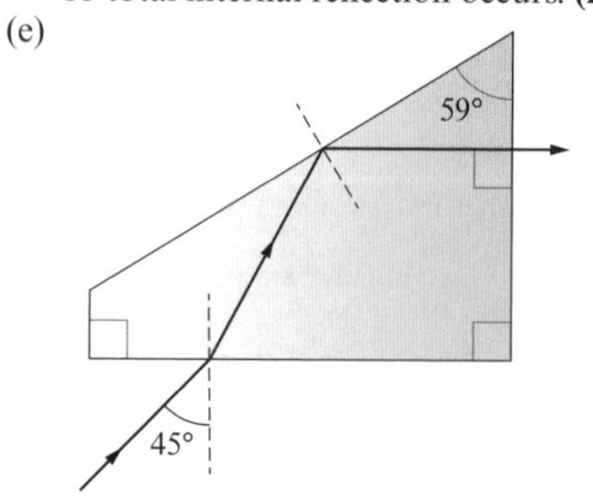

(2)

76. The photoelectric effect

1 (a) Electrons would not be emitted as photoelectrons if the zinc plate was positively charged or neutral, as a positively charged plate would attract the negatively charged photoelectrons and prevent emission. **(2)**

(b) The frequencies in visible light are lower than the threshold frequency for zinc, that is, electrons do not gain enough energy to escape from the surface, so no emission of photoelectrons can take place. **(2)**

(c) The photoelectric effect is only observed when incident light has a frequency greater than the threshold frequency. Increasing the intensity has no effect on the frequency, that is, the energy of individual photons, and so has no relevance to whether the effect is observed or not. **(2)**

(d) Ultraviolet light has a higher frequency than visible light. Its frequency is greater than the threshold frequency for zinc and so it does provide enough energy to eject photoelectrons and discharge the gold-leaf electroscope. **(2)**

(e) More UV photons per second means more photoelectrons are emitted per second, so a higher intensity of UV light will discharge the gold-leaf electroscope more rapidly. **(1)**

(f) Different metals require greater or lesser amounts of energy in order to eject photoelectrons, so different metals have different threshold frequencies. **(2)**

77. Einstein's photoelectric equation

1 (a) $E = hf = \frac{hc}{\lambda} = \frac{6.63 \times 10^{-34} \times 3.00 \times 10^{8}}{(589 \times 10^{-9})} = 3.38 \times 10^{-19}$ J **(2)**

(b) joules are converted to electronvolts by dividing by 1.60×10^{-19}, so $3.38 \times 10^{-19}\,\text{J} = \frac{3.38 \times 10^{-19}}{1.60 \times 10^{-19}} = 2.11$ eV **(1)**

2 The work function of a metal is the minimum amount of energy required to eject an electron from the surface of the metal. **(1)**

3 (a) work function $\phi = hf_0$, so $f_0 = \frac{\phi}{h} = \frac{2.36 \times 1.60 \times 10^{-19}}{(6.63 \times 10^{-34})} = 5.70 \times 10^{14}$ Hz. **(2)**

(b) maximum KE is the photon energy minus the work function $= hf - \phi = h(f - f_0) = 6.63 \times 10^{-34} \times \left[\frac{3.00 \times 10^{8}}{(465 \times 10^{-9})} - 5.70 \times 10^{14}\right] = 5.01 \times 10^{-20}$ J **(2)**

4 $E = hf = \frac{hc}{\lambda}$, so $\lambda = \frac{hc}{E}$

2.29 eV is $2.29 \times 1.60 \times 10^{-19} = 3.66 \times 10^{-19}$ J, so the longest wavelength that could result in the emission of photoelectrons from potassium is $\lambda = \frac{6.63 \times 10^{-34} \times 3.00 \times 10^{8}}{(3.66 \times 10^{-19})} = 5.43 \times 10^{-7}$ m or 543 nm, which is visible (green) light **(2)**

5 $f_0 = 6.90 \times 10^{14}$ Hz, so $\phi = 6.63 \times 10^{-34} \times 6.90 \times 10^{14} = 4.57 \times 10^{-19}$ J (2.86 eV) **(2)**

78. Determining the Planck constant

1 (a) The resistor limits the current in the LED, which would otherwise rise too rapidly once the LED starts to conduct. **(1)**

(b) $W = eV = 1.60 \times 10^{-19} \times 1.96 = 3.14 \times 10^{-19}$ J **(2)**

(c) photon energy = energy transferred, so $hf = eV_F$, and as $c = f\lambda$, $eV_F = \frac{hc}{\lambda}$, which can be rearranged to give $V_F = \frac{hc}{e\lambda}$ **(2)**

(d)

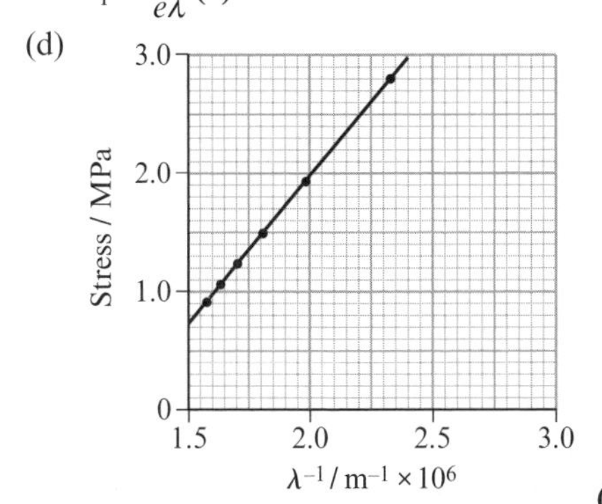

(5)

(e) gradient = $V_F\lambda = 1.24 \times 10^{-6}$ V m

$h = \text{gradient} \times \frac{e}{c} = \frac{1.24 \times 10^{-6} \times 1.60 \times 10^{-19}}{(3.00 \times 10^{8})} = 6.61 \times 10^{-34}$ J s **(2)**

79. Electron diffraction

1 (a) Diffraction occurs due to the superposition of waves and therefore if we see diffraction patterns produced by electrons passing through graphite, we must conclude that the electrons are behaving as waves. **(2)**

(b) Diffraction occurs because the wavelength of the electrons is comparable to the atomic spacing in the graphite. 'Polycrystalline' means made of very many crystals. These will have countless different orientations, leading to many overlapping diffraction patterns as electrons pass through the crystals. A single crystal would be expected to produce a diffraction pattern consisting of dots. However, as there are very many crystals with very many orientations, the patterns of dots have very many orientations and combine to form rings. **(2)**

(c) D – $\frac{p^2}{2m} = \frac{mv^2}{2}$ **(1)**

(d) The electrons of charge e are accelerated by a p.d. V and gain kinetic energy equal to eV.

$eV = \frac{1}{2}mv^2$ and $v = \sqrt{\frac{2eV}{m}} = \sqrt{\frac{2 \times 1.68 \times 10^{-19} \times 3000}{9.11 \times 10^{-31}}}$

$= 3.25 \times 10^{7}\,\text{m s}^{-1} \approx 3 \times 10^{7}\,\text{m s}^{-1}$ **(2)**

(e) the de Broglie wavelength of a particle is given by $\lambda = \frac{h}{p}$ where p is the momentum, mv, and h is the Planck constant, which is equal to 6.63×10^{-34} J s. $\lambda = \frac{h}{mv} = \frac{6.63 \times 10^{-34}}{(9.11 \times 10^{-31} \times 3.25 \times 10^{7})} = 2.24 \times 10^{-11}$ m **(1)**

(f) A – of smaller radius and brighter **(1)**

80. Wave–particle duality

1 (a) Young's double slit experiment provides evidence of the wave nature of light. Only waves exhibit superposition effects such as diffraction and interference, both of which are involved in the formation of fringes when light passes through a double slit. It is therefore appropriate to conclude from it that light travels as waves. **(3)**

(b) An experiment to show the photoelectric effect using a gold-leaf electroscope and zinc plate demonstrates the particle nature of light. The ability of light to discharge a charged gold-leaf electroscope depends on its frequency, in that only high-frequency light can discharge the gold-leaf electroscope. There is no explanation for this observation in a wave theory, so it must be concluded that light travels as discrete particles or quanta. **(3)**

2 As individual photons are being detected, this is strong evidence of the particle nature of light. The distribution of photons, however, maps out exactly the same pattern as would be predicted by the wave model. As interference is a wave phenomenon and photons are particles, we can only conclude that neither model tells the whole story and light may at times behave like a particle and at times behave like a wave. **(3)**

81. Exam skills

1 (a) The threshold frequency is the lowest frequency of light that will result in the emission of photoelectrons from a particular metal surface. **(2)**

(b)

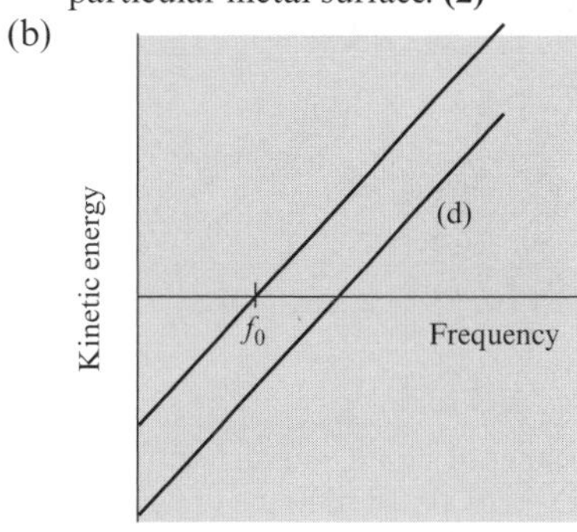

(1)

(c) The threshold frequency would be the frequency of photons that just have enough energy to knock electrons out of the metal surface but no more i.e. when $E_{max} = 0$ and $hf_o = \phi$. **(2)**

(d) see graph **(2)**

(e) A photon is a concentrated packet of electromagnetic radiation e.g. light or a quantum of energy of electromagnetic radiation. Its energy is given by $E = hf$. **(2)**

(f) All the energy of a photon is acquired by one particular photoelectron, but part of that energy is the work function, the minimum energy required to escape from the metal surface. If all the remaining energy is transferred into movement of the photoelectron, i.e. kinetic energy, it is the maximum, but in fact some of the energy might be transferred somewhere else e.g. to thermal energy in the metal. **(2)**

82. Temperature and thermal equilibrium

1 D **(1)**

2 The random motion of small visible particles as a result of thermal bombardment by much smaller, invisible molecules. **(2)**

3 (a) When two bodies are in thermal equilibrium, there is no net transfer of thermal energy between them, and they are at the same temperature. **(1)**

(b) Internal energy is the sum of the random distribution of kinetic and potential energies associated with the molecules of a system. **(1)**

(c) Temperature is a measure of how hot an object is, or in other words whether thermal energy will be transferred between two objects. **(1)** It is related to the amount of kinetic energy contributing to the object's internal energy. **(1)**

4 (a) (i) 332 K **(1)**
(ii) 90 K **(1)**
(b) 462 °C **(1)**

83. Solids, liquids and gases

1 C **(1)**

2 (a) In sections A, C and E there is no change of **state**. All the thermal energy supplied increases the **random kinetic energies** of the molecules, causing a **rise in temperature**. In sections B and D there is a change of **state**, so the thermal energy supplied is increasing the **potential energy** of the molecules as it breaks **bonds**, with no rise in **temperature**. **(4)**
(b) This region shows the time when ice is being converted into water at 0 °C. The energy supplied is changing the state of the water from solid (ice) to liquid (water). **(1)**
(c) This is the time when the energy being supplied is raising the temperature of the water. **(1)**

3 (a) Energy must be supplied to break bonds between molecules so that molecules close to the surface are able to escape. **(2)**
(b) Only the most energetic molecules have enough energy to escape, so the average molecular energy in the liquid that remains falls. Its temperature therefore falls. **(2)**

84. Specific heat capacity

1 $\Delta E = mc\Delta\theta = 25 \times 15 \times 2100 = 78\,750$ J **(2)**

2

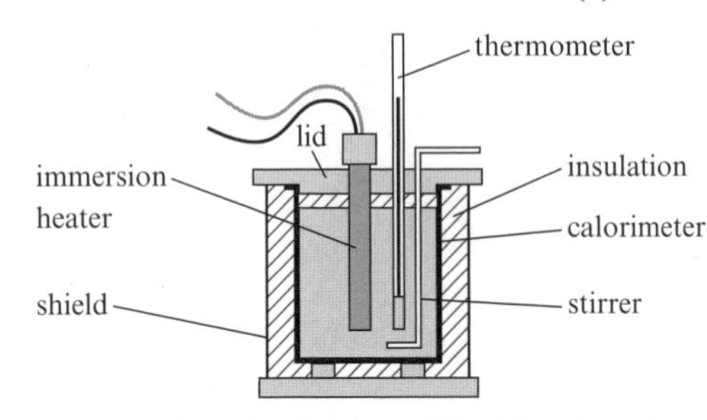

For example: using beaker of liquid and an electric immersion heater. Measure the mass of liquid used using a top-pan balance. Measure the current, voltage and time of heating using an ammeter, voltmeter and stopclock. Measure the temperature rise $\Delta\theta$ using a thermometer. Calculate the energy supplied using $E = IVt$. Calculate the specific heat capacity using $c = \frac{E}{m\Delta\theta}$.
Precautions: lag the beaker and use a lid; stir the liquid to ensure uniform temperature; start timing after heater has heated itself; start heating from below room temperature and increase to same amount above room temperature to reduce heat loss. **(8)**

3 energy transferred from copper to water $E = m_{Cu}c_{Cu}\Delta\theta_{Cu}$
$= m_W c_W \Delta\theta_w$
$0.250 \times 385 \times \Delta\theta_{Cu} = 2.0 \times 4200 \times \Delta\theta_w$
$\Delta\theta_{Cu} = 87.27 \times \Delta\theta_w$
if final temperature of water and copper $= T$, then $\Delta\theta_{Cu}$
$= 50 - T$ and $\Delta\theta_W = T - 20$
substitute into first equation:
$50 - T = 87.27 \times (T - 20)$
$50 + 1745.4 = 88.27T$
$\frac{1795.4}{88.27} = T$
$= 20.3$ °C **(4)**

85. Specific latent heat 1

1 D **(1)**

2 (a) When the ice melts it absorbs energy from the drink (the latent heat of fusion).
The water formed by the melted ice is at 0 °C and absorbs more energy from the drink, cooling it until all the liquid is at the same temperature. If you add water at 0 °C you only get the second cooling effect (which depends only on specific heat capacity and not on latent heat). **(3)**
(b) (i) $E = mL = 50 \times 10^{-3} \times 334 \times 10^3 = 16\,700$ J **(2)**
(ii) The water formed by the ice melting is colder than the orange squash, so more heat will flow from the orange squash until a uniform equilibrium temperature is reached. **(2)**

86. Specific latent heat 2

1 C **(1)**

2 energy needed to raise the temperature from 25 °C to 100 °C is $E = mc\Delta\theta = 1.20 \times 4200 \times 75 = 378\,000$ J **(1)**
additional energy needed to change it from water to steam at 100 °C is $E = mL = 1.20 \times 2.26 \times 10^6 = 2.71 \times 10^6$ **(1)**
total energy = 3 090 000 J **(1)**

3 (a) Electrical energy supplied to the heater changes the state of the liquid.
The amount of energy supplied can be calculated from $E = IVt$ using an ammeter, voltmeter and stopclock.
The vapour escapes through the holes in the flask and is condensed by the condenser and collected in the beaker.
This is weighed on a top-pan balance to find the mass of liquid that has changed state.
The equation to use is: $E = IVt = mL$ where L is the latent heat of vaporisation, $L = \frac{IVt}{m}$. **(6)**
(b) If timing begins before the liquid reaches its boiling point then some of the energy supplied goes to raising the temperature of the liquid rather than changing its state. This will result in an overestimate of the latent heat of vaporisation. **(2)**

87. The kinetic theory of gases

1 C **(1)**

2 (a) Pressure is force per unit area. Collisions between the container walls and the molecules result in a change of momentum of the molecules. An outward force equal but opposite to the average rate of change of momentum of the molecules is exerted on the walls. **(3)**
(b) When the gas is compressed into a smaller volume, the molecules, which still have the same mean and rms speed, collide more frequently with the walls, so the rate of change of linear momentum increases and the force increases too. So pressure increases. **(3)**

3 (a) Molecules of the vapour repeatedly collide with air molecules and follow a random path, so whilst they travel rapidly between collisions their average displacement from their starting point increases only slowly. **(2)**
(b) The rate of diffusion will depend on the rms speed of the molecules. From $pV = \frac{1}{3}NmC_{rms}$ we can see that $\sqrt{C_{rms}}$ is proportional to $\frac{1}{\sqrt{m}}$. We also know that $\frac{1}{3}mC_{rms} = \frac{3}{2}kT$ so $\sqrt{C_{rms}}$ is proportional to $\sqrt{T}$. So the rate of diffusion will be proportional to $\frac{\sqrt{T}}{\sqrt{m}}$. **(4)**

88. The gas laws: Boyle's law

1 (a) In this experiment the volume of the air is the dependent variable. We measure the length of the air column, which will be directly proportional to the volume as long as the cross-sectional area is constant ($V = \pi r^2 l$). **(2)**
(b) Temperature is a control variable in this experiment. If the gas is compressed quickly the temperature will rise, causing an additional increase in pressure and Boyle's law will not be followed. **(2)**
(c) A graph of pressure against $\frac{1}{l}$ (length l is proportional to V) will be a straight line through the origin if pressure is inversely proportional to volume. **(4)**

2 (a) $p_1V_1 = p_2V_2$
$1.2 \times 10^7 \times 0.0040 = 1.0 \times 10^5 \times V_2$
$V_2 = (1.2 \times 10^7) \times \frac{(0.0040)}{(1.0 \times 10^5)} = 0.48\,\text{m}^3$ **(2)**
(b) Oxygen acts as an ideal gas and obeys Boyle's law. There is no change in temperature. **(2)**

89. The gas laws: the pressure law

1 (a) Heat the water bath slowly, stirring continuously.
- Measure and record pressure and temperature (in degrees Celsius) over a range.
- Plot a graph of pressure against temperature. This should give a linear relationship.
- Extrapolate back – the intercept on the temperature axis is absolute zero. (A sketch graph would be a good addition to your answer; see Question 2.) **(5)**

(b) ΔT of 60 °C (60 K) gives Δp of 23 kPa
assuming linearity, 1 kPa reduction for each $\frac{60}{23\,°\text{C}}$
so pressure will be zero at $20 - \left(\frac{102 \times 60}{23}\right) = -246\,°\text{C}$ **(3)**

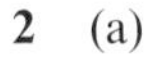
2 (a)

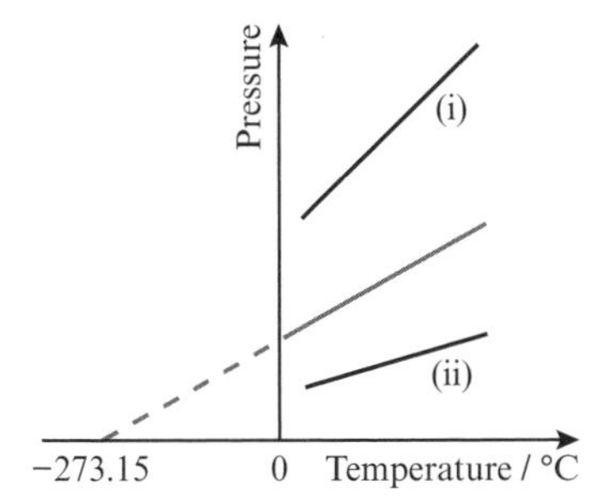

(i) double mass: a straight line with steeper gradient, which could be extended to absolute zero. **(1)**
(ii) half mass: a straight line with shallower gradient, which could be extended to absolute zero. **(1)**

(b) A real gas would condense to a liquid at some point so the pressure would drop below the dotted line. **(3)**

90. The equation of state of an ideal gas

1 C **(1)**
2 (a) steeper line starting at − 273.15 °C **(2)**
(b) shallower line starting at − 273.15 °C with all volume values half of original values.

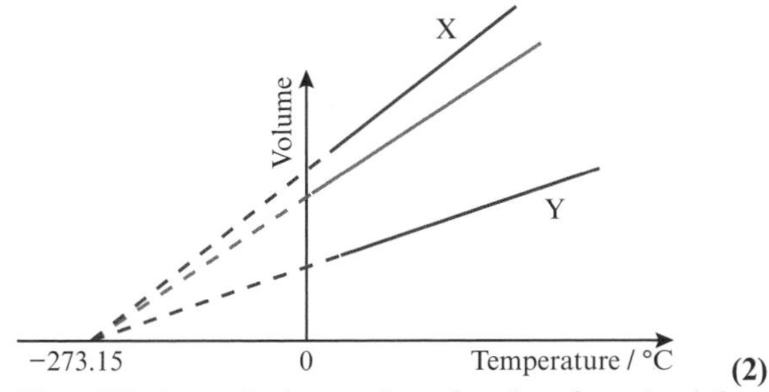

(2)

3 (a) $pV = nRT$ where n is the number of moles of gas (and thus directly proportional to the number of molecules) and R is a universal constant. Therefore, for the same values of p and T, V is always the same for a given value of n.
$V = n\left(\frac{RT}{p}\right)$ **(3)**
(b) By Avogadro's law there are twice as many **hydrogen atoms** as **oxygen atoms** involved in the reaction (twice the volume of **hydrogen**). The simplest way to combine the **atoms** is 2 **hydrogen** to 1 **oxygen** (H_2O). **(3)**

91. The kinetic theory equation

1 B **(1)**
2 (a) 441 m s^{-1} **(1)**
(b) 195 662 m^2 s^{-2} **(1)**
(c) 442 m s^{-1} **(1)**
3 T_2: at higher temperatures more molecules will have greater KE **(2)**

4 (a) mean KE $= \frac{3}{2}kT = \frac{3}{2} \times 1.38 \times 10^{-23} \times 313 = 6.48 \times 10^{-21}$ J **(2)**
(b) mean KE $= \frac{1}{2}mC_{\text{rms}} = \frac{3}{2}kT$
$C_{\text{rms}} = \frac{3kT}{m} = \frac{2 \times 6.48 \times 10^{-21}}{(4.7 \times 10^{-26})} = 2.75 \times 10^5$
$C_{\text{rms}} = 525\,\text{m s}^{-1}$ **(2)**
(c) At the same temperature, the molecules all have the same mean KE. **(2)**

92. The internal energy of a gas

1 D **(1)**
2 (a) $pV = nRT = NkT$
$n = \frac{pV}{RT} = \frac{110 \times 10^3 \times 3 \times 10^{-3}}{(8.31 \times 298)} = 0.133$ moles
number of moles $= nN_A = 8.02 \times 10^{22}$ molecules. **(4)**
(b) internal energy of n moles of an ideal gas is given by
$U = \frac{3}{2}nRT = \frac{3}{2}pV = \frac{3}{2} \times 330 = 495$ J **(2)**
3 (a) $\frac{3}{2}kT = \frac{3}{2} \times 1.38 \times 10^{-23} \times 303 = 6.3 \times 10^{-21}$ J **(2)**
(b) $\frac{3}{2}kT = \frac{1}{2}C_{\text{rms}}$
$C_{\text{rms}} = \frac{3kT}{m} = \frac{3 \times 1.38 \times 10^{-23} \times 303}{(4.7 \times 10^{-26})} = 267 \times 10^3$
$C_{\text{rms}} = 520\,\text{m s}^{-1}$ **(2)**

93. Exam skills

1 apparatus: sealed flask of gas, heat source, thermometer, pressure gauge
explanation: Place the sealed flask in a water bath. Heat the water bath and measure temperature and pressure over a range of temperatures.
measurements: temperature, pressure **(5)**
2 plot pressure against temperature in degrees Celsius extrapolate back to intercept on temperature axis **(2)**
3 Temperature is related to mean molecular kinetic energy. When the particles have zero kinetic energy that must be the lowest possible temperature. **(2)**
4 (a) It will be the same, because they are at the same temperature and mean kinetic energy $= \frac{3}{2}kT$. **(2)**
(b) $\frac{1}{2}mC_{\text{rms}} = \frac{3}{2}kT$
$C_{\text{rms}} = \frac{3kT}{m} = 3 \times 1.38 \times 10^{-23} \times \frac{(22 + 273)}{(2.7 \times 10^{-25})} = 45\,000$,
so $C_{\text{rms}} = 210\,\text{m s}^{-1}$ **(3)**

94. Angular velocity

1 C **(1)**
2 (a) $\frac{2\pi}{(3.16 \times 10^7)} = 1.99 \times 10^{-7}\,\text{rad s}^{-1}$ **(3)**
(b) $v = \frac{2\pi \times 150 \times 10^9}{(3.16 \times 10^7)} = 2.98 \times 10^4\,\text{m s}^{-1}$ **(1)**
(c) $v = \frac{2\pi \times 6400 \times 10^3}{(24 \times 3600)} = 470\,\text{m s}^{-1}$ **(2)**
(d) After one year the Earth will be back in its original position so the angle θ is just the angular displacement of Mars during one Earth year.
angular displacement $\theta = \omega\Delta T$
$\omega = \frac{2\pi}{T} = \frac{2\pi}{(1.87 \times 3.16 \times 10^7)}$
$\theta = \omega\Delta T = \frac{2\pi}{(1.87 \times 3.16 \times 10^7) \times (3.16 \times 10^7)} = 3.35$ rad,
192° **(3)**

95. Centripetal force and acceleration

1 (a) $F = \frac{mv^2}{r} = 8640$ N toward the centre of the circle. **(2)**
(b) The car's wheels are turned and the tyres exert an outward frictional force on the road. By Newton's third law the road produces an equal inward force on the car's tyres. **(2)**

(c) There is a limit to the maximum frictional force from the road. At higher speeds the required centripetal force is greater $\left(F = \frac{mv^2}{r}\right)$. If this is greater than the maximum frictional force then the car cannot turn in a circle of that radius at that speed and so it skids. **(2)**

2 (a) When the bob moves through its lowest position, it is moving in circular motion, so there must be a resultant force towards the centre of the circle (i.e. the point of suspension). The upward tension must therefore be greater than the downward weight in order to provide this resultant centripetal force. **(3)**

(b) resultant force = tension − weight = $\frac{mv^2}{r}$

tension = $mg + \frac{mv^2}{r}$ = 0.5886 + 0.11772 = 0.71 N **(4)**

96. Simple harmonic motion

1 Record the time for 10 oscillations. Repeat three times to obtain an average value of $10T$ and divide by 10 to find T. Ignore/repeat anomalous results. Use a fiducial marker placed at the centre of the oscillations to reduce timing errors.

Calculate f from $f = \frac{1}{T}$. **(5)**

2 (a) Force is directly proportional to displacement from equilibrium position. Force is directed towards equilibrium position. **(2)**

(b) The force is directed towards the equilibrium position and is directly proportional to displacement for initial displacements up to 9.0 mm. For small initial amplitudes the conditions for simple harmonic motion (s.h.m.) are met so it would undergo s.h.m. However, for larger initial displacements the restoring force is not directly proportional to displacement so the oscillations would not be simple harmonic. **(4)**

97. Solving the s.h.m. equation

1 A **(1)**

2 (a) $t = \frac{1}{f} = 0.50\,s$ **(1)**

(b) (i) $x = A\cos(2\pi ft) = 8.0\cos(2\pi \times 2.0 \times 0.125) = 0\,cm$ **(1)**

(ii) $x = 8.0\cos(2\pi \times 2.0 \times 0.25) = -8.0\,cm$ **(1)**

(iii) $x = 8.0\cos(2\pi \times 2.0 \times 0.40) = +2.5\,cm$ **(1)**

(c) $v = -A2\pi f\sin(2\pi ft)$ so max $= -A2\pi f = 1.01\,m\,s^{-1}$ as the oscillator passes its equilibrium position. **(2)**

(d) $F = -kx = -m\omega^2 x = -m(2\pi f)^2 x = 0.25 \times 4\pi^2 \times 2.0^2 \times 0.08$ = 3.2 N at maximum amplitude. **(2)**

(e) maximum KE (at maximum velocity)

$= \frac{1}{2} \times mv^2 = \frac{1}{2} \times 0.25 \times (1.01)^2 = 0.13\,J$

At this point the potential energy is 0 (equilibrium position) so total maximum energy = 0.13 J. **(2)**

3 $v = -\omega A\sin(2\pi ft)$ $\quad a = -\omega^2 A\cos(2\pi ft)$ **(2)**

98. Graphical treatment of s.h.m.

1 (a) 62–64 mm **(1)**

(b) 17–19 Hz **(1)**

(c) cosine curve with maximum velocity of about $7.3\,m\,s^{-1}$ **(4)**

(d) negative sine curve with maximum acceleration of about $805\,m\,s^{-2}$ **(4)**

99. Energy in s.h.m.

1 D **(1)**

2 (a) gravitational potential energy, kinetic energy, elastic potential energy: GPE→KE→EPE→KE→GPE **(2)**

(b) $x = \frac{-mg}{k} = \frac{-0.500 \times 9.81}{42} = 0.117\,m$ **(2)**

(c) (i) $E = \frac{1}{2}kx^2 = \frac{1}{2} \times 42 \times (0.117)^2 = 0.287\,J$ **(1)**

(ii) $E = \frac{1}{2} \times 42 \times (0.097)^2 = 0.197\,J$ **(1)**

(iii) $E = \frac{1}{2} \times 42 \times (0.137)^2 = 0.394\,J$ **(1)**

(d) $\Delta GPE = mg\Delta h = 0.500 \times 9.81 \times 0.04 = 0.196\,J$ $= 0.394 - 0.197\,J$ **(2)**

(e) max. KE = 0.197 J; $v = \sqrt{\left(\frac{2E}{m}\right)} = 0.888\,m\,s^{-1}$ **(3)**

(f) new amplitude $x_2 = \frac{1}{4}$ of x_1, and E is proportional to x^2, therefore fraction of E remaining is $\frac{1}{4^2} = \frac{1}{16}$. **(1)**

100. Forced oscillations and resonance

1 (a) (i) Natural frequency is the frequency of oscillations when the oscillator is displaced and released with no external forces acting on it. The forcing frequency is the frequency of the external forces (in this case from the vibration generator) acting on the system. **(2)**

(ii) $f = \frac{1}{(2\pi)} \times \sqrt{\left(\frac{k}{m}\right)} = \frac{1}{(2\pi)} \times \sqrt{\left(\frac{30}{0.065}\right)} = 3.42\,Hz$ **(1)**

(b) (i) curve starts at A_0, peaks at about 3.4 Hz, decays asymptotically toward zero. **(3)**

(ii) Resonance occurs when the driver (forcing) frequency is very close to the natural frequency of the driven oscillator and results in a large amplitude of response. **(2)**

(iii) At resonance, the driver continually transfers energy to the driven oscillator, causing its amplitude to grow. At the same time, work done by the oscillator against damping forces transfers energy to heat. The amplitude at resonance is determined by the point at which these two energy transfers occur at an equal rate. **(2)**

(c) Starts at A0; lower peak at resonance; curve lies below undamped curve; frequency of resonance slightly lower than natural frequency of oscillator. **(2)**

101. Exam skills

1 (a) (i) $f = \frac{1}{(2\pi)} \times \sqrt{\left(\frac{k}{m}\right)} = 1.48 \sim 1.5\,Hz$ **(3)**

(ii) $E = \frac{1}{2}kx^2 = \frac{1}{2} \times 6500 \times 0.084^2 = 23\,J$ **(2)**

(iii) maximum $a = -A(2\pi f)^2 = -0.084 \times 4\pi^2 \times (1.5)^2$ $= -7.5\,m\,s^{-2}$ **(2)**

(iv) Max acceleration = (7.3 + 9.81) so max force = ma = 60 × 17.1 = 1025 N ~ 1030 N **(2)**

(b) (i) If the bicycle continued to undergo s.h.m. after hitting the bump it would be very difficult to ride. The damper dissipates the energy of the oscillation so that the bike returns to normal after hitting the bump. **(2)**

(ii) 8.4 cm marked on compression axis at point where graph starts. Two cycles of oscillation with amplitude decaying to zero. **(4)**

102. Gravitational fields

1 B **(1)**

2 C **(1)**

3 Assume that Alan Shepard hit the ball with the same initial speed as on Earth.
The horizontal component of the ball's velocity will not be affected but the reduced value of g affects the vertical component. Initially the vertical component of velocity will be the same as on Earth but it will take longer for the Moon's gravity to reduce this to zero and then bring it back down. The time of flight (time to fall to the ground) increases by a factor of $\frac{9.81}{1.63}$, or 6.0 times. The horizontal component of the velocity is the same as on Earth so the range will be at least 6 times greater (1500 m). In addition to this there is no atmosphere on the Moon, so the ball's average speed will be greater. It can travel considerably further than 1500 m and so may have a range in excess of a mile (1600 m). **(4)**

4 (a) constant in magnitude and direction at all points **(1)**

(b) Over distances small compared with the radius of the Earth, the angle between field lines is so small that they can be treated as parallel and equally spaced. **(1)**

103. Newton's law of gravitation

1 (a) $F = \frac{Gm_1m_2}{r^2}$ so $\frac{F_A}{F_B} = \frac{Gm_1m_2}{r_A^2} \times \frac{r_B^2}{Gm_1m_2} = \frac{r_B^2}{r_A^2} = \frac{(249)^2}{(207)^2}$
$= 1.45$ **(2)**
(b) toward the Sun **(1)**
(c) speed is increasing and the radius of curvature of its path is getting smaller **(1)**
(d) At B, Mars is at its furthest point from the Sun, so it has its greatest gravitational potential energy. As it accelerates around its orbit the gravitational potential energy decreases (becomes more negative), transferring to kinetic energy. At A, the planet is closest to the Sun and has its maximum kinetic energy and least gravitational potential energy. As it moves along its orbit back towards B, kinetic energy is transferred back to gravitational potential energy. **(4)**
(e) at A **(1)**
(f) maximum gravitational force at A (closest point)
$F_A = \frac{Gm_1m_2}{r_A^2} = 6.67 \times 10^{-11} \times 6.39 \times 10^{23} \times$
$\frac{1.99 \times 10^{30}}{(207 \times 10^9)^2} = 1.98 \times 10^{21}\,N$ **(2)**

2 gravitational field strength $g = \frac{\text{gravitational force}}{\text{mass}} = \frac{F}{m}$

gravitational force on a mass m is $F = \frac{GMm}{r^2}$ where M is the mass of the Moon and r is its radius.

$g = \frac{F}{m} = \frac{GM}{r^2} = \frac{6.67 \times 10^{-11} \times 7.35 \times 10^{22}}{(1740 \times 10^3)^2} = 1.62\,N\,kg^{-1}$ **(2)**

104. Kepler's laws for planetary orbits

1 (a) An imaginary line between a planet and the Sun sweeps out equal areas during equal intervals of time. **(1)**
(b) By Kepler's second law, planets sweep out the same area in the same time interval when they are close to the Sun as when they are far away.
This means they must have smaller speeds when they are farther from the Sun, and so spend a longer period of time moving along this part of their orbit. **(2)**

2 $\frac{r^3}{T^2} = \frac{(150 \times 10^6)}{1^2}$, so:
Mercury: 0.24 years
Venus: 109 million km
Mars: 1.9 years **(3)**

3 Kepler's third law will apply to all satellites of the same central body.
This means that $\frac{r^3}{T^2}$ is the same for the satellite as for the Moon.
For the satellite to be geostationary its period must be $T = 24$ hours.
$\frac{r^3}{T^2} = \frac{(380\,000)^3}{(27.3 \times 24)^2}$ so $r_S^3 = \frac{(24)^2 \times (380\,000)^3}{(27.3 \times 24)^2}$
$r_S = 42\,000\,km$ **(3)**

105. Satellite orbits

1 D **(1)**

2 (a) gravitational forces provide the centripetal force:
$\frac{GMm}{r^2} = mr\omega^2$
and $\omega = \frac{2\pi}{T}$
$\frac{GMm}{r^2} = mr\left(\frac{2\pi}{T}\right)^2$
$T = \sqrt{\left(\frac{4\pi^2 r^3}{GM}\right)}$, which is independent of m. **(3)**
(b) $\frac{GM}{r^2} = r\left(\frac{2\pi}{T}\right)^2$
$r^3 = \frac{GMT^2}{(2\pi)^2} = \frac{(6.67 \times 10^{-11}) \times (6.0 \times 10^{24}) \times (2.5 \times 3600)^2}{4\pi^2}$
$r = 9.4 \times 10^6\,m$; radius of Earth is $6.400 \times 10^6\,m$
altitude $= 3.0 \times 10^6\,km$ **(3)**

3 (a) KE is maximum at A, closest to Earth, and falls as spacecraft moves from A to B. GPE rises from a minimum at A to a maximum at B. Then GPE falls as it moves from B to A and KE rises from B to A. From A to B, KE is being transferred to GPE and from B to A, GPE is being transferred to KE. The total energy is constant. **(3)**
(b) The departure point will not affect the amount of fuel because the total energy of the spacecraft is constant in the orbit. This means it needs the same additional energy from burning fuel to get to infinity (zero GPE) from any point on its orbit. **(4)**

106. Gravitational potential

1 (a) The gravitational potential at a point in space is the gravitational potential energy per unit mass at that point. **(2)**
(b) Gravitational potential is zero at infinite distance, and all gravitational forces are attractive. This means work must be done to move any mass from a point in space to infinity. If energy must be supplied to raise the potential to zero, then it must have started as a negative value. **(2)**
(c) Gravitational field strength is the negative gradient of the gravitational potential. The stronger the gravitational field strength, the more rapidly the potential changes with position. **(2)**
(d)

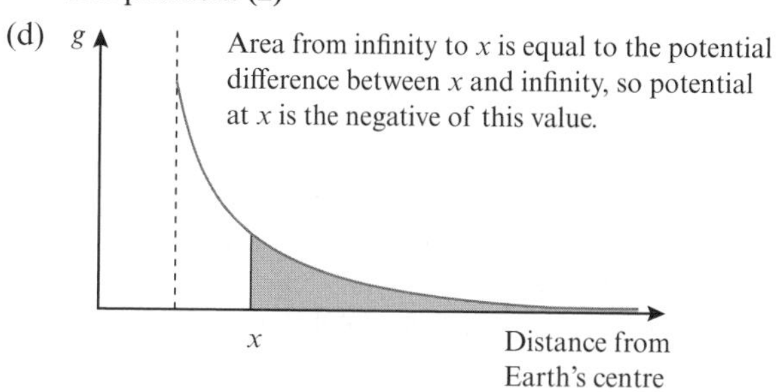

(3)
(e) The change in GPE of an object is equal to the change in gravitational potential between two points multiplied by the mass moved between those points. **(2)**

2 (a) equipotentials at $9.81\,J\,kg^{-1}$, $19.6\,J\,kg^{-1}$, $29.4\,J\,kg^{-1}$ etc. **(2)**
(b) vertical and horizontal distances are small compared with the radius of the Earth **(2)**

107. Gravitational potential energy and escape velocity

1 (a) the work done is equal to the change in gravitational potential energy $W = m\Delta V_g$
$\Delta V_g = -30\,MJ\,kg^{-1} - (-40\,MJ\,kg^{-1}) = +10\,MJ\,kg^{-1}$
$W = m\Delta V_g = 1200 \times 1.0 \times 10^7 = 1.2 \times 10^{10}\,J$, increase **(3)**
(b) (i) The minimum KE must be just enough to overcome GPE.
$\Delta V_g = -60\,MJ\,kg^{-1}$ so GPE $= -60 \times 10^6 \times 500$
$= -3.0 \times 10^{10}\,J$ **(2)**
(ii) KE $= \frac{1}{2}mv^2$ so escape velocity $v = \sqrt{\left(\frac{2 \times 3.0 \times 10^{10}}{500}\right)}$
$= 1.1 \times 10^4\,m\,s^{-1}$ ($11\,km\,s^{-1}$) **(3)**
(iii) atmospheric friction dissipates energy **(2)**

108. Exam skills

1 The gravitational force of attraction between two point masses m_1 and m_2 a distance d apart is directly proportional to the product of their masses and inversely proportional to the square of their separation: $F \propto \frac{m_1m_2}{d^2}$. **(1)**

2 gravitational force per unit mass at a point in space, $N\,kg^{-1}$ **(2)**

3 gravitational force = centripetal force, $F = \frac{Gm_1m_2}{d^2} = md\omega^2$
where $\omega = \frac{2\pi}{T}$
orbital radius $d^3 = \frac{Gm_ET^2}{4\pi^2}$
$= \frac{6.67 \times 10^{-11} \times 6.0 \times 10^{24} \times (2 \times 3600)^2}{4\pi^2} = 5.26 \times 10^{20}$
$d = 8.070 \times 10^6$
altitude $= d - 6.400 \times 10^6 = 1.7 \times 10^6\,m$ (1700 km) **(4)**

4 (a) work that must be done per unit mass to move a mass from infinity (zero potential) to the point in space. **(2)**
(b) gravitational potential energy (GPE) at infinity is zero. Work must be done against attractive forces to move a mass from a point in space to infinity, so the GPE at the initial point must be less than at infinity, i.e. negative. **(3)**

5 (a) in 1 year $\Delta GPE = \frac{-GMm}{r_2} - \left(\frac{-GMm}{r_1}\right)$

$= -6.67 \times 10^{-11} \times 6.0 \times 10^{24} \times 7.3 \times 10^{22} \times \left(\frac{r_1}{r_1 r_2} - \frac{r_2}{r_1 r_2}\right)$

$r_1 r_2$ is approximately r^2, so ΔGPE

$= \frac{-2.92 \times 10^{37} \times -38 \times 10^{-3}}{(3.9 \times 10^8)^2} = \frac{7.3 \times 10^{18}\,J}{year^{-1}} = 230\,GW$ **(3)**

(b) Moon's KE reduces, Earth's rotational KE reduces **(2)**

109. Formation of stars

1 C **(1)**

2 (a) As the core became hotter it reached the condition for nuclear fusion reactions to occur. This created an outward radiation pressure that has balanced the inward pressure of gravitational collapse. **(3)**
(b) Nuclear fusion reactions in the core of the Sun are converting hydrogen to helium so the percentage of hydrogen will fall and the percentage of helium will rise. (After about 5 billion years the Sun will have used up much of its hydrogen and will eventually expand into a red giant.) **(3)**

3 Any three from:
- planets are less massive than stars
- stars are sources of high-energy EM radiation
- planets orbit stars
- planets do not have nuclear fusion reactions in their cores
- planets are visible by reflection, stars are visible by radiation. **(3)**

4 Conditions at the core of a more massive star are **more extreme** – hotter and higher pressure than in a less massive star. This greatly increases the rate of **nuclear fusion** reactions, reducing the lifetime of the star. **(2)**

110. Evolution of stars

1 C **(1)**

2 (a) This is the maximum **mass** for a white dwarf star and is about **1.4** times the mass of the Sun. More massive stars will collapse beyond the **white dwarf** stage to form **neutron stars** or even **black holes.** **(4)**
(b) electron degeneracy pressure **(1)**
(c) neutron degeneracy pressure **(1)**

3 (a) expanding region of hot gases, white dwarf at centre **(2)**
(b) core of red giant (later stage for a low- to medium-mass star) continues to collapse and its temperature rises. Once temperature reaches 10^8 K, helium in the core fuses rapidly (helium flash) to give carbon and oxygen.
Most of the material around the core is ejected as a cloud of gas; this is the planetary nebula.
(The high-density remnant of the core glows brightly at first, but no further fusion takes place. This is a white dwarf.) **(4)**

111. End points of stars

1 B **(1)**

2 (a) (i) red giant **(1)**
(ii) Once most of the hydrogen in the core has **fused** into helium, the radiation pressure **decreases**. The star contracts under **gravitational** forces, and helium fusion begins in the star's outer layers. The energy released in the helium shell causes radiation pressure that makes the outer layer of the star **expand again**. As the surface area of the star increases, its surface temperature **decreases.** **(3)**
(b) a planetary nebula **(1)**
(c) The Sun's core remnant will be less than the Chandrasekhar limit. **(3)**

3 when the mass of the core remnant is so great that gravitational collapse cannot be prevented by electron or neutron degeneracy pressure **(2)**

112. The Hertzsprung–Russell diagram

1 (a) Stellar luminosity: total power radiated by a star **(2)**
(b) White dwarf star: end stage of the stellar lifecycle for a star like our Sun. When fuel runs out the core collapses under its own gravitational field but stops when it has a very high density. The surface temperature is very high. **(2)**
(c) Neutron star: end stage of the stellar lifecycle for a star more massive than our Sun. When fuel runs out the core collapses under its own gravitational field and atoms are crushed to form a dense ball of neutrons. **(2)**
(d) Black hole: end stage of the stellar lifecycle for a star much more massive than our Sun. When fuel runs out the core collapses under its own gravitational field and the collapse continues without limit. The resulting object is called a black hole because light cannot escape from its gravitational field. **(2)**
(e) Main-sequence star: Once a star has formed and nuclear fusion reactions are taking place in its core it can remain like this for billions of years. During this stage it is said to be on the main sequence. When main-sequence stars are plotted on the Hertzsprung–Russell diagram they form a diagonal band. **(2)**

2 Stage 1: gas cloud collapses to form a protostar and nuclear fusion reactions begin in the core
Stage 2: star reaches equilibrium and remains on the main sequence for a long time
Stage 3: nuclear fuel begins to run out, star begins to collapse but then restarts fusion and expands to become a red giant
Stage 4: outer layers of red giant drift away leaving a white-hot core, a white dwarf star, which gradually cools down **(6)**

113. Energy levels in atoms

1 B **(1)**

2 (a) Electrons are bound to the **nucleus** so work has to be done to **separate them** from the nucleus. This raises their energy to **zero** so the initial potential energy must have been negative. **(2)**
(b) $E = \frac{-13.6\,eV}{n^2}$ so for $n = 1$ to $n = 3$, $\Delta E = -13.6 \times \left(\frac{1}{3^2} - \frac{1}{1^2}\right)$
$= 12.1\,eV$ **(3)**
(c) $E = hf = \frac{hc}{\lambda}$ so $\lambda = \frac{hc}{E}$
$E = 12.1\,eV = 1.9 \times 10^{-18}\,J$
$\lambda = \frac{6.63 \times 10^{-34} \times 3.00 \times 10^8}{1.9 \times 10^{-18}} = 1.03 \times 10^{-7}\,m$ **(3)**
(d) $\Delta E = -13.6 \times \left(\frac{1}{\infty^2} - \frac{1}{1^2}\right) = 13.6\,eV$ **(2)**
(e) An elastic collision conserves momentum and kinetic energy so no kinetic energy can be transferred to electron potential energy in the collision. This means that the collision energy must be below the energy needed to promote an electron from the $n = 1$ to $n = 2$ state in either atom (i.e. 10.2 eV). **(3)**

114. Emission and absorption spectra

1 B **(1)**

2 (a) Dark lines cross the continuous spectra, indicating that light energy has been absorbed at particular wavelengths. **(2)**
(b) Photons excite electrons in low energy levels inside an atom and cause them to jump to higher energy levels. The photon energy that can do this is equal to the energy jump that is allowed in the atom. When the electrons fall back to their original energy levels they re-emit the photons in random directions, so the intensity of the original light beam is reduced. **(3)**

3 $d\sin\theta = n\lambda$ so for first-order maximum ($n = 1$)
$\lambda = (d\sin\theta) = \frac{1}{400} \times 10^{-3} \times \sin(13.627)$
$= 5.89 \times 10^{-7}$ m (589 nm)
$\frac{1}{400} \times 10^{-3} \times \sin(13.650) = 5.90 \times 10^{-7}$ m (590 nm) **(4)**

115. Wien's law and Stefan's laws

1 (a) about 0.105 cm. **(1)**
(b) $T = \frac{2.9 \times 10^{-3}}{(0.105 \times 10^{-2})} = 2.76$, so about 2.8 K. **(2)**
(c) microwave **(1)**
(d) The matter in the early Universe must have been hot and thus emitted high-frequency (short wavelength) radiation. As spacetime expanded, the wavelength of that radiation would have been stretched, until now it is in the microwave region. **(2)**

2 (a) Both curves are the shape of black body radiation curves. The curve for Sirius is always above the curve for Betelgeuse and has a peak at a lower wavelength. The peak wavelength for Sirius is about 3×10^{-7} m. The peak wavelength for Betelgeuse is about 8×10^{-7} m. **(5)**
(b) Total radiated power is the same as luminosity.
Power radiated is given by Stefan's law:
$L = 4\pi r^2 \sigma T^4 = 4\pi \times (8.2 \times 10^{11})^2 \times 5.67 \times 10^{-8} \times (3500)^4$
$= 7.2 \times 10^{31}$ W **(2)**

116. The distances to stars

1 (a) If the distances to the nearest stars are very large then the parallax angles are small. It was therefore impossible for ancient astronomers to detect the very small parallax angles with the naked eye, and telescopes had not been invented. **(2)**
(b) The further the distance from Earth, the smaller the parallax angle. The parallax angle is given by $p = \frac{1}{d}$ where p is the parallax in arcseconds of arc and d is the distance in parsec. **(2)**
(c) $p = \frac{1}{d}$ so $d = \frac{1}{0.314} = 3.18$ pc $= 9.87 \times 10^{16}$ m $= 10.4$ ly **(4)**
(d) (i) 100 ly = 30.6 pc
from Earth: $p = \frac{1}{30.6} = 0.03$ arcseconds
$= 1.58 \times 10^{-7}$ rad
from Hipparcus: $p = \frac{1}{306} = 0.003$ arcseconds
$= 1.58 \times 10^{-8}$ rad **(2)**
(ii) assuming stars are evenly distributed throughout the volume of sphere centred on the Earth (and on Hipparcus), ratio of volumes = ratio of numbers of stars $\frac{V_E}{V_H} = \frac{\frac{4}{3}\pi r_E^3}{\frac{4}{3}\pi r_H^3} = \frac{100^3}{1000^3}$
$= \frac{1}{1000}$ **(3)**

117. The Doppler effect

1 (a) As the car moves from A to C the component of its velocity directed toward or away from the spectator changes. This causes a Doppler shift, affecting the received frequency. **(2)**
(b) The Doppler effect depends on the relative velocity of the source of sound to the person hearing the sound. Since the engine and driver are both moving at the same velocity, there will be no change in the frequency of the sound. **(2)**
(c) (i) From A to B the frequency heard by the spectator is above the dotted line and falling at increasing rate. At B the frequency heard by the spectator is equal to that heard by the driver. From B to C the frequency heard by the spectator is below the dotted line and falling at decreasing rate. **(3)**
(ii) From A to B there is a component of velocity toward the spectator so the Doppler shift increases the received frequency.
At B there is momentarily no component of velocity directed toward the spectator so the frequency is not Doppler shifted.
From B to C the component of velocity is away from the spectator so the Doppler shift reduces the frequency. **(3)**

2 The wavelength is redshifted, so the stars are moving away from the Earth.
Doppler shift $z = \frac{\Delta\lambda}{\lambda} = \frac{v}{c}$, so speed of recession =
$3.00 \times 10^8 \times \frac{(21.132 - 21.106)}{21.106} = 3.7 \times 10^5$ m s^{-1} **(2)**

118. Hubble's law

1 C **(1)**

2 (a) Red shift $z = \frac{\Delta\lambda}{\lambda_0}$ is related to recession velocity v by the equation $z = \frac{v}{c}$.
Recession velocity is related to distance by Hubble's law, $v = H_0 d$, so we can combine these two equations to give $zc = H_0 d$ and rearrange this to obtain $d = \frac{zc}{H_0}$. The additional information required is the speed of light and the Hubble constant. **(4)**
(b) The red shift would be affected by any additional (local) motion of the galaxy on top of cosmological expansion. The value of the Hubble constant has some uncertainty. **(2)**

3 (a) Hubble's law is $v = H_0 d$, where v is recession velocity of a galaxy at distance d and H_0 is the Hubble constant. **(3)**
(b) The Hubble time is $\frac{1}{H_0} = 13.7$ billion years. When the age is converted to seconds, $H_0 = 2.3 \times 10^{-18}$ s^{-1}. **(2)**

119. The evolution of the Universe

1 D **(1)**

2 (a) All galaxies rotate. The centripetal force that keeps stars in orbit around the galactic nucleus is provided by gravity. However, the total gravitational attraction from all the visible matter is not enough to provide the required force so dark matter must provide the additional gravitational force. **(4)**
(b) Two of:
- It interacts via the gravitational force.
- It does not emit electromagnetic radiation (is dark).
- It is non-baryonic matter. **(2)**

3 (a) recession velocity is directly proportional to distance: $v = H_0 d$ **(1)**
(b) 71 km s^{-1} Mpc^{-1} $= \frac{71 \times 10^3 \times 1}{3.1 \times 10^{-16} \times 10^{-6}\ \text{m s}^{-1}\text{m}^{-1}}$
$= 2.29 \times 10^{-18}$ s^{-1} **(2)**
(c) $v = H_0 d = 71$ km s^{-1} Mpc^{-1} $\times$ 100 Mpc = 7100 km s^{-1} **(2)**
(d) distance 100 Mpc = $100 \times 3.1 \times 10^{19}$ km
galaxies moving apart at 7100 km s^{-1}, so were together $\frac{3.1 \times 10^{21}}{7100} = 4.37 \times 10^{17}$ s ago, that is, 13.8×10^9 years
This is an estimate of the time since the Big Bang, or the age of the Universe. **(2)**

120. Capacitors

1 C **(1)**

2 (a) $C = \frac{Q}{V} = 50\,\mu$F **(2)**
(b) (i) energy = area under graph **(1)**
(ii) area of a triangle = $\frac{1}{2} \times$ base $\times$ height
energy $= \frac{1}{2}QV = \frac{1}{2} \times 40 \times 2000 \times 10^{-6} = 4.0 \times 10^{-2}$ J **(2)**
(iii) energy $= \frac{1}{2}QV = \frac{1}{2} \times 20 \times 1000 \times 10^{-6} = 1.0 \times 10^{-2}$ J **(2)**

3 (a) $Q = CV = 2.64\,\text{mC}$ **(2)**

(b) $I = \frac{V}{R} = 0.12\,\text{A}$ through both ammeters **(2)**

(c) As the capacitor charges up the potential difference across it increases. This opposes the e.m.f. of the battery, so the potential difference across the resistor (which is the difference of these two values) falls so current decreases. **(2)**

121. Series and parallel capacitor combinations

1 D **(1)**

2 (a) $Q = CV = 22 \times 10^{-6} \times 6.0 = 132 \times 10^{-6}\,\text{C}$ ($132\,\mu\text{C}$) **(2)**

(b) $282\,\mu\text{C}$ **(2)**

(c) The p.d. across each capacitor and the cell is 6.0 V so there is no p.d. across the resistor and no current will flow. **(2)**

(d) $414\,\mu\text{C}$, 6.0 V **(2)**

(e) total capacitance in parallel is $C_1 + C_2 = 22 + 47 = 69\,\mu\text{F}$ and $C = \frac{Q}{V} = \frac{414}{6.0} = 69\,\mu\text{F}$ **(2)**

(f) total capacitance in series C: $\frac{1}{C} = \frac{1}{C_1} + \frac{1}{C_2} = 0.067\,\mu\text{F}^{-1}$
$C = 15\,\mu\text{F}$ **(2)**

122. Capacitor circuits

1 (a) The time constant determines how long it takes for a capacitor to charge or discharge: over one time constant the charge falls to approximately 37% of its initial value. **(3)**

(b) units of RC are $\Omega\text{F} = \text{VA}^{-1}\text{CV}^{-1} = \text{A}^{-1}\text{C} = \text{C}^{-1}\text{sC} = \text{s}$ **(2)**

2 (a) $RC = 0.022\,\text{s}$ **(2)**

(b) $(0.37)^5 = 0.007$, that is, in five time constants a discharged capacitor will be 99.3% charged, so $5 \times CR = 0.11\,\text{s}$. **(2)**

(c) Increasing C increases the final charge ($Q = CV$) so it takes longer to supply this charge. Increasing R decreases the charging current so again it takes longer to supply the charge. **(4)**

3 time constant $CR = 50 \times 10^{-6} \times 20 \times 10^3 = 1\,\text{s}$; $5 \times CR = 5\,\text{s}$ **(2)**

123. Exponential processes

1 D **(1)**

2 (a) Charging curve rising to almost 6 V in about 2.5 s then constant at 6 V to 3 s. Discharging curve reaching almost zero at about 5.5 s and then remaining close to zero. **(6)**

(b) $Q = CV = 6.0\,\text{mC}$ **(1)**

(c) Charging current I is equal to $\frac{V_R}{R}$, where V_R is the potential difference across the resistor. $V_R = 6.0 - V_C$ so will be maximum when the capacitor is uncharged ($V_C = 0$).
at $t = 0$, $I = \frac{V_R}{R} = 12\,\text{mA}$ **(2)**

(d) $Q = Q_0\,e^{-t/RC}$ so when charge is halved, $\frac{Q}{Q_0} = e^{-t/RC} = 0.5$

$\ln 0.5 = \frac{-t}{RC}$

$t = RC\ln 2 = 0.35\,\text{s}$ after it starts to discharge (3.35 s on graph) **(3)**

(e) Inverse of graph for capacitor so that p.d. across resistor and p.d. across capacitor add to 6.0 V at all times. **(2)**

124. Exam skills

(a) Time constant for the charging circuit is 4.0 s. Charging is 95% complete after 3 time constants so the voltage across the capacitor is well above 60 V and the flash will work. **(3)**

(b) (i) $W = \frac{1}{2}CV^2 = 0.014\,\text{J}$ **(2)**

(ii) $W = 0.0036\,\text{J}$ **(1)**

(c) $\frac{V}{V_0} = e^{-t/RC} = 0.5$ so $\frac{t}{RC} = \ln 2$ and $t = 1.4\,\text{ms}$ **(2)**

(d) energy released = 0.014 J in 1.4 ms, giving an average power of 10 W **(4)**

(e) The flash lamp stops conducting when the p.d. across it falls below 60 V, so the supply only has to recharge the capacitor from 60 V to 120 V. This takes less time than charging from 0 V. **(2)**

125. Electric fields

1 A **(1)**

2

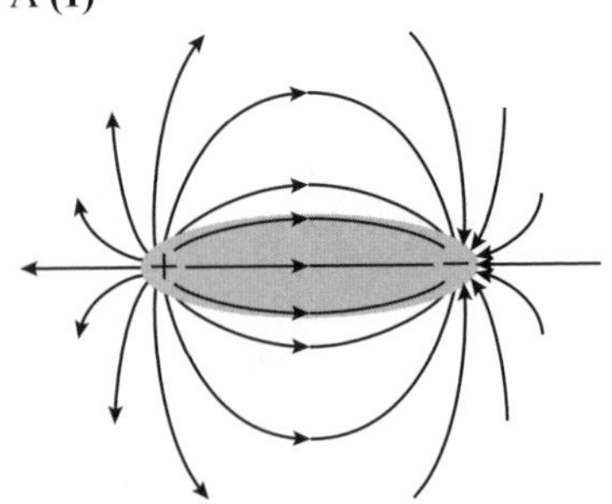

(3)

3 Field lines represent the force per unit charge that would be exerted if a small positive charge was placed at that point. If the field lines crossed they would represent two forces in different directions. These can always be resolved into a single field line. **(2)**

4 A **(1)**

126. Coulomb's law

1 C (because $F = \frac{Q_1Q_2}{4\pi\varepsilon_0 d^2}$ so both Q and d are multiplied by 4) **(1)**

2 (a) Taking forces to the right as positive, the resultant force on the electron is (by Coulomb's law) $F = \frac{Qe}{(4\pi\varepsilon_0 r^2)} - \frac{(-Qe)}{(4\pi\varepsilon_0 r^2)}$
$= 2 \times 8.0 \times 10^{-20} \times \frac{(-1.6 \times 10^{-19})}{(4\pi\varepsilon_0 r^2)}$
$= -3.7 \times 10^{-9}\,\text{N}$ (attraction)
The negative sign indicates a force to the left. **(4)**

(b) resultant force = force due to $-Q$ + force due to $+Q$:
$F = \frac{-Qe}{(4\pi\varepsilon_0 r^2)} + \frac{Qe}{[4\pi\varepsilon_0(3r)^2]} = 1.6 \times 10^{-9}\,\text{N}$ to right (repulsion) **(4)**

(c) The external field will exert a force on each end of the dipole.
The charges at each end of the dipole are opposite so the forces will be in opposite directions.
Unless the molecule is already aligned with the field these two forces will create a turning effect. As a result the molecules might align with the field so that the entire material becomes a dipole. **(4)**

127. Similarities between electric and gravitational fields

1 B **(1)**

2 gravitational $F = \frac{Gm_1m_2}{r^2}$
$= \frac{6.67 \times 10^{-11} \times 6.64 \times 10^{-27} \times 3.27 \times 10^{-25}}{r^2} = \frac{1.45 \times 10^{-61}}{r^2}$

electrostatic $F = \frac{Q_1Q_2}{(4\pi\varepsilon_0 r^2)} = \frac{3.20 \times 10^{-19} \times 1.27 \times 10^{-17}}{(4 \times \pi \times 8.85 \times 10^{-12} \times r^2)}$
$= \frac{3.65 \times 10^{-26}}{r^2}$

The electrostatic force is more than 10^{35} times greater than the gravitational attraction so Rutherford was justified in ignoring gravitational effects. **(4)**

3 (a) For example, two positive charges placed close to one another. Work has to be done to put them into this configuration because they exert repulsive forces on each other. If work is done, their electrostatic potential energy must increase from zero (the electrostatic potential energy at infinity). **(3)**

(b) All gravitational forces are attractive, so the field does work to pull them together from infinity, and work would have to be done to separate them. This means that the gravitational potential energy (GPE) at infinity (zero) is greater than that of any configuration of masses so they must always have a negative GPE. **(2)**

128. Uniform electric fields

1 C **(1)**

2 (a) $V = 3 \times 10^6 \times 2.5 \times 10^{-3} = 7500\,\text{V}$ **(2)**

(b) (i) $E = \frac{V}{d} = \frac{500}{(4 \times 10^{-3})} = 125\,\text{kV m}^{-1}$ **(1)**

(ii) $E = \frac{500}{(8 \times 10^{-3})} = 62.5\,\text{kV m}^{-1}$ **(1)**

(iii)

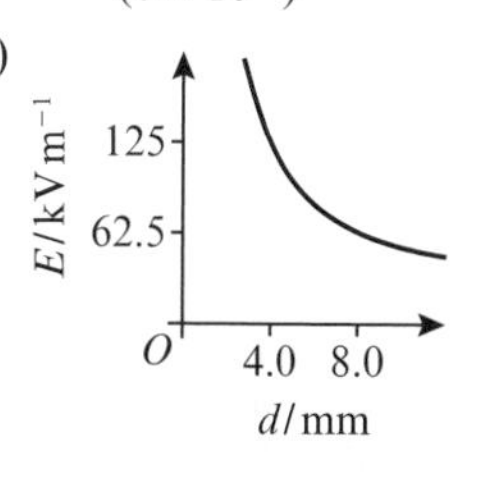

(3)

129. Charged particles in uniform electric fields

1 (a) B **(1)**

(b) $t = \frac{d}{v}$ where v where v is the initial velocity and thus the horizontal component of the velocity throughout **(1)**

(c) $a = \frac{eE}{m}$, vertically downwards **(2)**

(d) initial vertical velocity component = 0, so vertical displacement by the time the electron has passed through horizontal distance d is $s = \frac{1}{2}\left(\frac{eE}{m}\right)\left(\frac{d}{v}\right)^2$ **(2)**

(e) $w = \left(\frac{eE}{m}\right)\left(\frac{d}{v}\right) = \frac{eEd}{mv}$ **(2)**

(f) We need the direction of the velocity vector as the electron leaves the field.
horizontal component = v
vertical component = $w = \frac{eEd}{mv}$
$\tan\theta = \frac{w}{v} = \frac{eEd}{mv^2}$ so $\theta = \tan^{-1}\left(\frac{eEd}{mv^2}\right)$ **(3)**

130. Electric potential and electric potential energy

1 A **(1)**

2 (a) $V = \frac{Q}{(4\pi\varepsilon_0 r)}$ **(1)**

(b) (i) At 10 cm from centre, $160 \times 10^3 = \frac{Q}{(4\pi\varepsilon_0 \times 0.1)}$ so
$Q = 4\pi\varepsilon_0 \times 16 \times 10^3$
then at 20 cm (i.e. 10 cm from dome surface)
$V = \frac{Q}{(4\pi\varepsilon_0 \times 0.2)} = \frac{16 \times 10^3}{0.2} = 80\,\text{kV}$ **(1)**

(ii) $V = \frac{16 \times 10^3}{0.4} = 40\,\text{kV}$ **(1)**

(c) Line starts at 10 cm, 160 kV and goes through 20 cm, 80 kV and 40 cm, 40 kV in a smooth curve. **(4)**

3 electric potential energy change on separation to infinity
$= \frac{Qq}{(4\pi\varepsilon_0 r)}$
$= \frac{(46 \times 1.60 \times 10^{-19})^2}{(4\pi \times 8.85 \times 10^{-12} \times 5 \times 10^{-15})} = 9.7 \times 10^{-11}\,\text{J} = 610\,\text{MeV}$ **(3)**

131. Capacitance of an isolated sphere

1 D **(1)**

2 (a) $Q = V \times 4\pi\varepsilon_0 R = 5000 \times 4\pi \times 8.85 \times 10^{-12} \times 0.02$
$= 1.1 \times 10^{-8}\,\text{C}\ (11\,\text{nC})$ **(2)**

(b) (i) $V = \frac{Q}{(4\pi\varepsilon_0 r)}$ so
$V_A - V_B = \frac{Qq}{(4\pi\varepsilon_0 \times 0.1)} - \frac{Qq}{(4\pi\varepsilon_0 \times 0.15)} = \frac{Qq}{(4\pi\varepsilon_0 r)} \times 3.33$
substitute expression for Q in 2(a) and cancel constants: $V_A - V_B = 5000 \times 0.02 \times 3.33 = 330\,\text{V}$ **(4)**

(ii) $Q(V_A - V_B) = 2.0 \times 10^{-9} \times 330 = 660 \times 10^{-9} = 0.66\,\mu\text{J}$ **(4)**

3 (a) $C = 4\pi\varepsilon_0 R = 4\pi \times 8.85 \times 10^{-12} \times 0.05 = 5.6 \times 10^{-12} = 5.6\,\text{pF}$ **(2)**

(b) energy $= \frac{1}{2}V^2C = 0.5 \times (2000)^2 \times 5.6 \times 10^{-12} = 1.1 \times 10^{-5}\,\text{J}$ (always 11 μJ) **(2)**

(c) $I = \frac{V}{R} = \frac{2000}{(22 \times 10^3)} = 0.091\,\text{A}$ **(2)**

(d) $RC = 22 \times 10^3 \times 5.6 \times 10^{-12} = 0.12\,\mu\text{s} << 50\,\text{ms}$ so the capacitor would be fully discharged. **(4)**

132. Representing magnetic fields

1 D **(1)**

2 (a) A hard magnetic material is difficult to **magnetise** or **demagnetise**.
A soft magnetic material is easy to **magnetise** and **demagnetise**. **(2)**

(b) Soft iron strengthens the field. It is easily magnetised and demagnetised so the electromagnet, and the lifting force, can be switched on and off rapidly. **(4)**

3 something like:

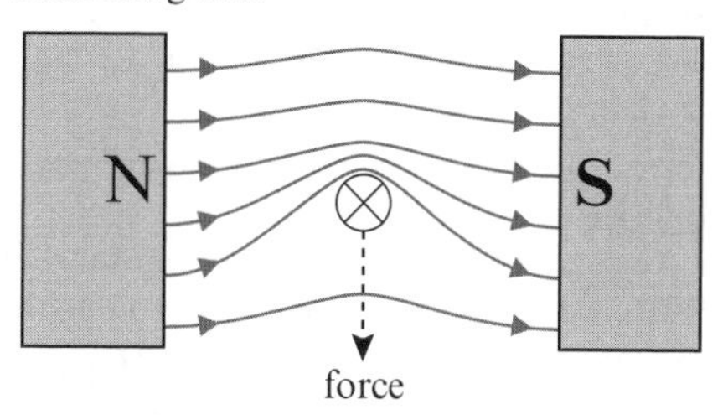

(4)

4 diagram which:

- represents Earth's field by uniform field lines
- shows compass needle and poles
- forces on each pole (N with force parallel and S with force antiparallel to external field)
- discusses turning effect and equilibrium condition when needle lines up with field. **(4)**

133. Force on a current-carrying conductor

1 C **(1)**

2 (a) arrow pointing downwards **(1)**

(b) The force on the magnet will be equal and opposite to the force on the wire (by Newton's third law).
The force on the magnet therefore acts in an **upward** direction.
This force will **decrease** the contact force of the magnet on the balance so the reading on the balance will **decrease**. **(3)**

(c) (i) $F = mg = 0.61 \times 10^{-3} \times 9.81 = 6.0 \times 10^{-3}\,\text{N}\ (6.0\,\text{mN})$ **(2)**

(ii) $F = BIl\sin\theta$
$B = \frac{F}{Il\sin\theta} = \frac{6.0 \times 10^{-3}}{(2.0 \times 0.060 \times 1)} = 0.050\,\text{T}$ **(2)**

134. Charged particles in a region of electric and magnetic field

1 B **(1)**

2 (a) arc of a circle with centre of curvature below point of entry to field **(2)**

(b) as in (a) but larger radius of curvature ($r = \frac{mv}{Bq}$ and v is larger) **(2)**

3 for a charge Q and mass M moving at speed v perpendicular to a magnetic field of strength B, $\frac{Mv^2}{r}$
$= BQv$, so $r = \frac{Mv}{BQ}$
for U-235, $r_{235} = \frac{235v\text{u}}{Bq}$
for U-238, $r_{238} = \frac{238v\text{u}}{Bq}$
the separation after a semicircle will be $2r_{238} - 2r_{235}$
$= 2 \times (238 - 235)\frac{v\text{u}}{Bq} = \frac{6v\text{u}}{Bq}$ **(4)**

135. Magnetic flux and magnetic flux linkage

1 A **(1)**

2 The vertical component of the magnetic field strength is $B_{vert} = 36 \times 10^{-6} \times \sin 30°$.
The flux is given by $\varphi = B_{vert}A = 36 \times 10^{-6} \times \sin 30° \times 2.0 = 3.6 \times 10^{-5}$ Wb. **(3)**

3 (a) $\varphi = BA = 0.22 \times \sin 55° \times 0.0080 = 1.4 \times 10^{-3}$ Wb **(2)**
(b) $NBA = 50 \times 1.4 \times 10^{-3} = 72 \times 10^{-3}$ Wb.turn **(1)**

4 One cycle of a sine curve. Peak values indicated at $\pm NBA$. **(3)**

136. Faraday's law of electromagnetic induction and Lenz's law

1 (a)
- As the magnet moves up and down, its magnetic field creates a changing flux through the coil.
- Induced e.m.f. in the coil is **directly proportional to the rate of change** of flux linkage (Faraday's law).
- In this example, the switch **is open so there is not a complete circuit**, therefore there is **no current**. **(3)**

(b)
- When the switch is closed the induced e.m.f. will cause current to flow in the coil.
- The direction of current flow will oppose the change that caused it (Lenz's law).
- The currents will make the coil into an electromagnet, which repels the approaching magnet and attracts the receding magnet, damping the oscillations. **(3)**

(c) The moving magnet has to do work against the forces from the coil. This transfers mechanical energy to electrical energy in the coil. Energy is conserved. **(2)**

2 (a) change in flux linkage $= NB\Delta A = 50 \times 20 \times 10^{-3} \times 25 \times 10^{-4} = 2.5 \times 10^{-3}$ Wb.turn **(2)**
(b) $\varepsilon = \frac{-d(N\Phi)}{dt}$ so average $\varepsilon = \frac{2.5 \times 10^{-3}}{0.20} = 0.0125$ V **(3)**

137. The search coil

1 D **(1)**

2 (a) (i) 0 Wb.turn (ii) $N\Phi = NBA = 2.4 \times 10^{-3}$ Wb.turn
(iii) 0 Wb.turn (iv) -2.4×10^{-3} Wb.turn **(4)**
(v) Graph y-axis scaled -2.5×10^{-3} to $+2.5 \times 10^{-3}$ and unit is Wb. Shape is a sine curve with one cycle up to $t = T$. **(4)**
(b) (i) Shape is a negative cosine with one cycle up to $t = T$. **(2)**
(ii) **Faraday's** law says that **induced e.m.f.** is proportional to negative rate of change of **flux linkage** so the values on the e.m.f. graph are the **negative gradients** of the flux-linkage graph at the same time. **(3)**

138. The ideal transformer

1 (a) $V = \frac{120}{200} \times 300 = 180$ V **(2)**
(b) I in secondary circuit $= \frac{V}{R} = 3$ A
$\frac{n_s}{n_p} = \frac{I_p}{I_s}$ so $I_p = 3 \times \frac{300}{200} = 4.5$ A **(3)**
(c) The changing flux linkage in the soft iron core induces eddy currents that transfer electrical energy to heat and reduce the efficiency of the transformer. Laminating the core interrupts these eddy currents and reduces the losses, increasing the efficiency of the transformer. **(3)**

2 When the switch is closed the current in the primary coil increases from zero. This causes an increase in the magnetic flux in the iron core. This changing flux links the secondary coil and induces a voltage across it (Faraday's law). There is a complete circuit so a current flows and this creates a magnetic field in the coil surrounding the compass. The compass deflects to align with this field. Once the current has reached a steady value there is no longer a change of flux linkage so the induced e.m.f. falls to zero and so does the induced current. The compass needle now returns to its original position. **(8)**

139. Exam skills

1 (a) 50 Hz **(2)**
(b)
- Current in first coil creates a magnetic field through the coil.
- The current is alternating so it creates an alternating magnetic flux.
- The alternating flux links the second coil.
- By Faraday's law there is an induced e.m.f. proportional to the rate of change of magnetic flux linkage through the second coil. **(4)**

(c) 0 ms, 10 ms, 20 ms, 30 ms etc....
E.m.f. is directly proportional to rate of change of flux, and flux is directly proportional to the current in the first coil. The maximum values of induced e.m.f. will be at times when the current through the first coil has its maximum rate of change (where the gradient of the graph is steepest). **(3)**
(d) No energy is transferred. The second coil is an open circuit so no current flows. Power requires both e.m.f. and current ($P = IV$). **(2)**

140. The nuclear atom

1 (a) In a vacuum the alpha particles will not be deflected or stopped by colliding with air molecules. **(1)**
(b) Gold can be beaten into a very thin foil. **(1)**
(c) Most of the alpha particles passed through the gold foil with little **or no deflection even though the foil was hundreds of atoms thick**. This suggests that they did not get close to **any other particles**, so most of the space inside the atom must be **empty**. **(2)**
(d) (i) Alpha particles are known to carry a positive charge. A force is required to deflect them. Rutherford assumed that this force was an electrostatic force between the positive charge of the alpha particle and a small volume of concentrated charge – the nucleus. **(2)**
(ii) The proportion of alpha particles scattered through large angles was very small. This suggested that the chance of passing close to a nucleus was very low so the nucleus must occupy a very small fraction of the space inside the atom. **(2)**
(e) lowest alpha particle approaches closest to nucleus and is then deflected back along same path
middle alpha particle deflects upwards
top alpha particle deflects least and crosses path of middle alpha particle. **(3)**
(f) Since the nucleus is much smaller than the atom and yet contains most of its mass, it must have a much higher density than ordinary matter. **(2)**
(g) 79 protons, 197 − 79 = 118 neutrons **(2)**

141. Nuclear forces

1 (a) $\log r$: 0.425, 0.535, 0.638, 0.740, 0.823
$\log A$: 1.08, 1.45, 1.75, 2.08, 2.32 **(3)**
(b) $\log r = \log r_0 + n \log A$
If $\log r$ is plotted against $\log A$, the graph should be a straight line with gradient n and intercept $\log r_0$. **(3)**
(c) $n = 0.33$
$r_0 = 1.1 \times 10^{-15}$ m. Values determined from plotted graph. **(5)**

142. The Standard Model

1 C **(1)**

2 C **(1)**

3 A **(1)**

4 (a) strong nuclear force **(1)**
(b) (i) anti-up, anti-down, anti-down **(1)**
(ii) anti-up, anti-up, anti-down **(1)**
(c) three of: same mass; same magnitude of charge; same spin; opposite sign of charge **(3)**
(d) up and down quarks, and electrons **(3)**

143. The quark model of hadrons

1 B (1)

2 (a) Both neutrons and protons are baryons consisting of three quarks.
Neutrons are udd and protons **are uud**, so when a neutron decays to a proton, a **down** quark must change to an **up** quark. **(2)**

(b) The neutron is neutral and the proton has charge $+e$, so another particle must be emitted in the decay and this particle must have charge $-e$. **(2)**

3 (a) (i) mesons
(ii) zero
(iii) π^0 particle: up and anti-up or down and anti-down
(iv) π^+ particle: up and anti-down **(4)**

(b) The creation of the pion requires energy at least as great as the rest energy of the pion itself (or the rest mass $\times c^2$). This comes from the initial kinetic energy of the incoming proton in the proton beam. **(2)**

144. β^- decay and β^+ decay in the quark model

1 D (1)

2 (a) The beta-particle has lepton number +1 so a particle with lepton number − 1 must be emitted in order to conserve lepton number. However, charge is already conserved so the emitted particle must be a neutral anti-lepton: the anti-neutrino. This particle is also necessary to conserve momentum and energy. **(2)**

(b) one extra proton and one less neutron. **(2)**

(c) (i) baryon/baryon/lepton/lepton **(4)**
(ii) A neutron changes to a proton in the nucleus – both are baryons so there is no change in baryon number (electrons and neutrinos are not baryons). **(1)**
(iii) A neutron consists of **udd** quarks. In beta-minus decay, a neutron decays to become a **proton**. Since protons consist of **uud** quarks, a **down** quark must have changed to an **up** quark in the process. This is a change of flavour. **(2)**

145. Radioactivity

1 (a) The range of an ionising radiation depends on how strongly it interacts with matter. Gamma rays are **uncharged** so they are only **weakly** ionising and therefore they lose energy slowly as they move through the air. Alpha particles and beta particles are both **more strongly** ionising so they lose energy **faster** and have **shorter** ranges. **(2)**

(b) Alpha particles are strongly ionising but have very short range. They can be stopped by the outer layers of (mainly) dead skin. However, if they are ingested they can cause great damage to growing and dividing cells. **(2)**

2 (a) Radioactivity is a random process. **(1)**

(b) 81.2, 75.8 **(2)**

(c) Beta – there is a small reduction when thick card is used. If the source was an alpha source this would have fallen back to background. Beta and gamma can penetrate this (although a few beta particles would be absorbed). When lead is used the count rate falls back to background levels suggesting that all the radiation from the source has been absorbed. If it was a gamma emitter these readings would be higher. **(4)**

146. Balancing nuclear transformation equations

1 (a) (i) ${}^{234}_{91}\text{Th} \rightarrow {}^{234}_{91}\text{Pa} + {}^{0}_{-1}\text{e} + {}^{0}_{0}\bar{\nu}$ **(3)**
(ii) ${}^{230}_{90}\text{Th} \rightarrow {}^{226}_{88}\text{Ra} + {}^{4}_{2}\alpha$ **(3)**

(b) (i) change in baryon number is 232 − 208 = 24 therefore 6 alpha particles emitted **(1)**
(ii) 12 protons lost in alpha particles, but change in proton number is 90 − 82 = 8 protons. Therefore 4 neutrons in nucleus decay to protons, therefore 4 beta particles emitted. **(1)**

2 (a) The effect of beta-minus decay is to increase Z by 1 and decrease N by 1. This creates a new nucleus closer to (or on) the line of stability. **(3)**

(b) The effect of beta-plus decay is to decrease Z by 1 and increase N by 1. This creates a new nucleus closer to (or on) the line of stability. **(3)**

147. Radioactive decay 1

1 A **(1)**

2 (a) 215, 159, 116, 88, 69, 54, 45, 38, 36 **(2)**
(b) correctly plotted graph **(3)**
(c) about 28 cpm (judged from flattening of graph) **(1)**
(d) About 55 s. Working should show an attempt to find the half-life from more than one starting point. **(3)**

148. Radioactive decay 2

1 (a) The gamma rays are emitted randomly in all directions so **the detector only records those emitted in one direction.**
The efficiency of the detector is **not 100% – the detector only records a fraction of the gamma rays that pass through it.**
The body of the sample itself **may absorb some of the gamma rays inside it.** **(3)**

(b) (i) after one half-life $A = \frac{186}{2} = 93\,\text{kBq}$ **(1)**
(ii) after three half-lives $A = \frac{186}{2^3} = 23\,\text{kBq}$ **(2)**

(c) (i) $A = A_0 e^{-\lambda t}$ so after one half-life (t = 5.3 years) $1 = 2e^{-\lambda t}$
$\ln 0.5 = -\lambda t_{\frac{1}{2}}$
$$\lambda = \frac{-\ln 0.5}{(5.3 \times 365 \times 24 \times 60 \times 60)} = 4.1 \times 10^{-9}\,\text{s}^{-1}$$ **(1)**
(ii) $A = A_0 e^{-\lambda t}$ so when $A = \frac{A_0}{10}$ then $\frac{1}{10} = e^{-\lambda t}$
taking natural logs, $\ln 0.1 = \ln(e^{-\lambda t}) = -\lambda t$
$$\frac{-2.30}{(4.1 \times 10^{-9})} = -t$$
$t = 5.56 \times 10^8$ s, so every 17.6 years **(3)**

149. Einstein's mass – energy equation

1 (a) (i) $m = \frac{E}{c^2} = \frac{1.43 \times 10^8}{(3.00 \times 10^8)^2} = 1.6 \times 10^{-9}\,\text{kg}$ **(2)**
(ii) This is about one billionth of the mass of the reacting particles and so is negligible. **(2)**

(b) (i) energy per kg helium produced $E = 6.8 \times 10^{14}$ J
equivalent mass change per kg helium produced
$= m = \frac{E}{c^2} = \frac{6.8 \times 10^{14}}{(3.00 \times 10^8)^2} = 7.56 \times 10^{-3}\,\text{kg}$
percentage change = 0.8% **(2)**
(ii) A change in mass of 0.8% is measureable and not small enough to neglect. **(1)**

(c) matter–antimatter annihilation is 100% efficient so
(i) $\frac{1}{(1.6 \times 10^{-9})} = 6.3 \times 10^8$ times more efficient than a chemical reaction
(ii) $\frac{1}{(0.008)} = 130$ times more efficient than nuclear fusion. **(4)**

150. Binding energy and binding energy per nucleon

1 C **(1)**

2 mass of 8 protons and 8 neutrons = 16.127 528 u; mass deficit of nucleus Δm = 0.132 613 u; total binding energy $= c^2\Delta m = 1.99 \times 10^{-11}\,\text{J} = 125\,\text{MeV}$, binding energy per nucleon $= 7.79\,\text{MeV/nucleon} = 1.25 \times 10^{-12}\,\text{J/nucleon}$ **(4)**

3 (a) Iron-56 has the greatest value of binding energy per nucleon so it takes more energy per nucleon to split the nucleus up into individual nucleons than any other nucleus. **(2)**

(b) All three of these nuclides are peaks on the binding energy per nucleon curve so they are relatively more stable than the nuclides adjacent to them. This means that, when nuclei form, these will be more likely to form and so are produced in larger numbers (by nuclear fusion reactions in stars). **(3)**

(c) roughly 0.5 MeV/nucleon increase in binding energy for 12 nucleons so there will be approximately 6 MeV released per carbon-12 nucleus formed **(2)**

151. Nuclear fission

1 B **(1)**

2 (a) Main features: axes labelled, binding energy per nucleon on y-axis and $\frac{\text{nucleon}}{\text{mass number}}$ on x-axis; steep rise to peak at iron-56; shallow drop to uranium at end. **(3)**

(b) Splitting a heavy nucleus (e.g. uranium) can form two lighter nuclei. These have greater binding energy per nucleon so the energy difference can be released in the nuclear fission process. **(3)**

3 $a = 1, b = 0, c = 141, d = 56$. **(4)**

4 $N = \frac{1.0}{235} \times 6.02 \times 10^{23}$ fissions

total energy $= \frac{3.20 \times 10^{-11} \times 1.0}{235 \times 6.02 \times 10^{23}} = 8.2 \times 10^{10}$ J **(3)**

152. Nuclear fusion and nuclear waste

1 (a) Two nuclei have to approach one another close enough for the **strong nuclear** force to overcome the **coulomb (electrostatic)** repulsion, since both nuclei are positively charged. The **strong** nuclear force can then bind them together to form a heavier nucleus with the release of **energy**. This is very difficult to achieve in practice because the individual nuclei must have extremely high **energies** to get so close. This means creating and controlling plasmas at extremely **high temperatures**. **(4)**

(b) (i) charge: the proton numbers (the lower ones) represent the charges on the nuclei, and the sum is the same on each side (1 + 1 = 1 + 1)
baryon number: the baryon numbers (numbers of nucleons) on the top are also equal on each side (2 + 2 = 3 + 1) **(2)**

(ii) mass deficit $= 2 \times 2.013553 - (1.007276 + 3.015500)$
$= 4.33 \times 10^{-3}$ u $= 7.23 \times 10^{-30}$ kg
energy $= mc^2 = 7.23 \times 10^{-30} \times (3.00 \times 10^8)^2$
$= 6.51 \times 10^{-13}$ J, 4.07 MeV **(5)**

(iii) number of hydrogen and deuterium nuclei in 1.0 kg water $= \frac{2 \times 1.0}{0.018 \times 6.02 \times 10^{23}}$ number of deuterium nuclei only $= \frac{2 \times 1.0}{0.018 \times 6.02} \times \frac{10^{23}}{4500} = 1.49 \times 10^{22}$
maximum energy released $= 1.49 \times 10^{22} \times 6.51 \times 10^{-13}$
$= 9.68 \times 10^9$ J **(3)**

153. Exam skills

1 (a) $F = \frac{qV}{d}$ down the page **(2)**

(b) (i) into the page **(1)**

(ii) the force from the magnetic field is $F = B_1qv$ so $B_1 = \frac{F}{qv}$ but for the particles that continue in a straight line (those with velocity v) this force is cancelled out by the force
from the electric field $F = \frac{qV}{d}$
by substitution $B_1 = \frac{V}{vd}$ **(2)**

(iii) The force from the magnetic field depends on v but the force from the electric field does not, so the forces no longer balance. If $B > B_1$ the particles will be deflected upwards so they will not pass through the collimating slit. If $B < B_1$ the particles will be deflected downwards so they also will not pass through the collimating slit. **(2)**

(c) (i) the more massive isotope (Q) – the radius of curvature is
$r = \frac{mv}{Bq}$ but B, q and v are all the same for both isotopes so the radius is proportional to m **(2)**

(ii) $XY = 2 \times \Delta r = 2\left(\frac{m_2v}{Bq} - \frac{m_1v}{Bq}\right) = \frac{2v(m_2 - m_1)}{Bq}$ **(3)**

154. Production of X-ray photons

1 C **(1)**

2 (a) (i) X-ray photons gain energy from the kinetic energy of the incoming electrons. This means that each electron must have enough energy to emit an X-ray photon. The electron energy is given by eV. X-ray photons have large energy so eV must be large too. **(2)**

(ii) The process is very inefficient and a lot of heat is generated as the electrons collide with target. Copper is a very good thermal conductor so it helps to transfer heat away from the target and prevents it melting. Flowing water acts as a coolant into which the heat is transferred for removal. **(2)**

(iii) This prevents the electrons from scattering off air molecules. **(1)**

(b) maximum possible energy of photons is $eV = \frac{hc}{\lambda}$ so
minimum $\lambda = \frac{hc}{eV} = \frac{(6.63 \times 10^{-34} \times 3.00 \times 10^8)}{(1.6 \times 10^{-19} \times 40 \times 10^3)}$
$= 3.1 \times 10^{-11}$ m **(3)**

3 Incoming electrons eject atomic electrons in the target from **inner** energy levels. These lines are formed when outer electrons in target atoms make quantum jumps down into the lower **energy levels** and emit a **photon** of equivalent energy (and thus defined wavelength). The K_α line has a shorter wavelength so corresponds to a larger energy jump than the K_β **line**. **(3)**

155. X-ray attenuation mechanisms

1 B **(1)**

2 (a) Since the intensity decays exponentially it will have a constant half-**thickness**, so the intensity halves every **4.5** cm. 18 cm corresponds to four **half**-thicknesses so the intensity will be reduced to $\frac{1}{2^n}$ where n = **4**. The transmitted intensity is therefore $\frac{1}{16}$**, or 6.25 % (2)**

(b) $I = I_0 e^{-\mu x}$ so $\log I = \log I_0 - \mu x$
$\log 0.5 = \log 1 - \mu \times 0.045$ so $\mu = 15.4\,\text{m}^{-1}$ **(3)**

(c) $0.01 = \log 1 - 15.4 \times x$
$x = 30$ cm **(3)**

3 (a) mass of electron–positron pair is $2m_e = 2 \times 9.1 \times 10^{-31}$ kg
energy equivalent $E = mc^2 = 2 \times 9.1 \times 10^{-31} \times (3.00 \times 10^8)^2$
$= 1.64 \times 10^{-13}$ J **(2)**

(b) $E = \frac{hc}{\lambda}$ so $\lambda_{max} = 1.2 \times 10^{-12}$ m, therefore X-rays with a wavelength of 1.0×10^{-12} m can cause pair production. **(3)**

156. X-ray imaging and CAT scanning

1 B **(1)**

2 A **(1)**

3 (a) The brightest parts of the image are the **bones**. These actually **absorb** more of the X-rays than the **soft tissue**, so if the image brightness were proportional to the X-rays transmitted then these areas would be darkest. **(2)**

(b) The attenuation coefficient for bone is much greater than for the soft tissue because it has a higher density. **(2)**

4 Advantages of CAT scan: better resolution in the image; no need for patient to use a contrast medium (e.g. barium meal). Disadvantages: higher dose of X-rays increases risk of damage to cells; more complicated procedure; more expensive. **(4)**

157. Medical tracers

1 C **(1)**

2 (a) ${}^{99}_{42}\text{Mo} \rightarrow {}^{99}_{43}\text{Tc} + {}^{0}_{-1}\beta + {}^{0}_{0}\bar{\nu}$ **(2)**

(b) The relatively long **half-life** means that only **a little of the molybdenum** has decayed during transportation so the technetium generator is still useful when it arrives at the hospital. **(2)**

(c) Gamma-rays are penetrating so they can leave the body and be detected using a gamma-camera. Having a half-life of 6 hours is long enough to carry out diagnostic tests but short enough so that the activity inside the body drops to a low level in just a few days. **(3)**

3 (a) Beta-plus decay results in emission of a positron (antielectron). This will annihilate almost immediately with a nearby electron to create a pair of gamma-ray photons travelling in opposite directions. These leave the body and can be detected using a gamma camera. **(3)**

(b) Advantages: only a small quantity needs to be used; body gets rid of it quickly.
Disadvantage: it needs to be generated locally and used quickly. **(2)**

158. The gamma camera and diagnosis

1 C **(1)**

2 (a) scintillator: made of sodium iodide – absorbs gamma-ray photons and emits visible light **(2)**

(b) photomultiplier tubes: detect individual photons and generate a measureable electric current. **(2)**

3 (a) Electrons are negatively charged so will be accelerated by the potential difference between electrodes. When they strike each electrode they will have enough energy to eject several more electrons. **(2)**

(b) Each photon ejects one electron from the photocathode. This is multiplied five times at each of 12 anodes so the final number of electrons is **5** to the power **12** and the total charge is $\mathbf{5^{12}} \times 1.6 \times 10^{-19} = \mathbf{3.9 \times 10^{-11}}$ **C**. **(3)**

159. PET-scanning and diagnosis

1 B **(1)**

2 (a) The two gamma rays are created at the same instant and take a very short time to cross the ring. **(2)**

(b) If an event occurs closer to one side of the ring **than the other**, the gamma-ray photon will reach that side first. The difference in **arrival time** can be used to calculate the difference in **distances travelled** by each **photon** so it is possible to calculate where the original event was along a line running from **one detector** on the ring to the one opposite it. **(2)**

(c) (i) $ct = 0.5 \times 3.00 \times 10^8 \times 200 \times 10^{-12} = 300 \times 10^{-4}$ m (30 mm) **(3)**

(ii) The smaller the time differences that can be measured, the smaller the differences in distance travelled by the gamma-ray photons that can be measured, so smaller scale features can be imaged. **(3)**

160. Ultrasound

1 C **(1)**

2 B **(1)**

3 (a) time of one-way trip $t = \frac{0.048}{1400} = 3.4 \times 10^{-5}$ s so return trip time is 69 μs. **(2)**

(b) The shortest time between emitting a pulse and detecting its echo is for boundaries at **2.0 cm from the transmitter**. The time for the return trip of a pulse to this boundary is $\frac{2 \times 0.02}{1400} = 29$ μs. In order that the returning pulse cannot be confused with the emitted pulse, the maximum pulse duration must be **less than 29 μs**. **(3)**

(c) For ultrasound to have a wavelength of 0.50 mm the frequency must be greater than 2.8 MHz. This suggests that the waves themselves are capable of resolving details this small (for frequencies 2.8–10 MHz). However, additional factors will limit the ability to resolve detail (e.g. scattering, pulse duration, contrast). **(4)**

161. Acoustic impedance and the Doppler effect

1 A **(1)**

2 (a) Z = density × speed of sound for a particular medium. **(1)**

(b) The ratio of reflected intensity to incident intensity is given by the expression:

$$\frac{I_r}{I_0} = \left(\frac{Z_2 - Z_1}{Z_2 + Z_1}\right)^2.$$

This ratio will be larger if the difference in acoustic impedances between the two media is larger. **(2)**

(c)

Medium	Speed / m s^{-1}	Density / kg m^{-3}	Z / kg m^{-2} s^{-1}
Air	330	0.0012	**0.396**
Water	1480	1000	1.5×10^6
Bone	4080	**1910**	7.8×10^6
Muscle	**1590**	1070	1.7×10^6
Blood	1550	1060	$\mathbf{1.6 \times 10^6}$

(4)

(d) 0.09 % **(2)**

3 The fractional change in frequency is given by:

$$\frac{\Delta f}{f} = 2 \times \frac{v\cos\theta}{c}$$

This will be a maximum when $\cos\theta = 1$.

The percentage change is $\frac{\Delta f}{f} \times 100 = 2 \times \frac{0.40}{1500} \times 100 = 0.05$ %. **(3)**

162. Ultrasound A-scan and B-scan

1 A **(1)**

2 (a) LH pulse is reflection from front surface and RH pulse is reflection from back surface. **(2)**

(b) time difference = 4.2 × 25 μs = 105 μs, distance across object $= \frac{1500 \times 105 \times 10^{-6}}{2} = 0.079$ m (79 mm). **(3)**

(c) The spread of the pulses is an indication of uncertainty in time. Here, this uncertainty is about **10** μs. This corresponds to an uncertainty in total distance of **0.015** m and an uncertainty in the measurement of about **0.0075** m. **(3)**

(d) Two from:
- attenuation in the medium – the pulse from the back surface has further to travel.
- spreading of pulse – the longer the path the more it spreads.
- scattering from objects inside the medium. **(2)**

(e) B-scans provide information about the amplitude of the reflected pulse as well as the time it takes to return. This results in a recognisable image on the screen rather than a set of pulses. B-scans are built up from a series of static scans that are added together, whereas an A-scan is a single static scan. **(3)**

For your own notes

For your own notes

Published by Pearson Education Limited, 80 Strand, London, WC2R 0RL.

www.pearsonschoolsandfecolleges.co.uk

Copies of official specifications for all OCR qualifications may be found on the OCR website: www.ocr.org.uk

Typeset by Kamae
Produced by Out of House Publishing
Illustrated by Tech-Set Ltd, Gateshead

First published 2016

19 18 17 16

10 9 8 7 6 5 4 3 2 1

British Library Cataloguing in Publication Data
A catalogue record for this book is available from the British Library

ISBN 978 1 447 98435 1

Printed in Slovakia by Neografia

Acknowledgements
The publisher would like to thank the following for their kind permission to reproduce their photograph:

Shutterstock.com: a. v. ley 110, Jim Barber 156

All other images © Pearson Education